D0024991

Synthetic-Aperture Radar and Electronic Warfare

Walter W. Goj

Artech House
Boston • London

For a complete listing of the *Artech House Radar Library*,
turn to the back of this book

Synthetic-Aperture Radar and Electronic Warfare

Library of Congress Cataloging-in-Publication Data

Goj, Walter W.
Synthetic aperture radar and electronic warfare / Walter W. Goj.

p. cm.
Includes bibliographical references and index.
ISBN 0-89006-556-7
1. Radar—Military applications. 2. Synthetic aperture radar. 3. Radar—interference.
4. Electronic countermeasures. I. Title.

UG612.G65 1993 92-21739
623.7'348—dc20 CIP

685 Canton Street
Norwood, MA 02062

International Standard Book Number: 0-89006-566-7
Library of Congress Catalog Card Number: 92-21739

10 9 8 7 6 5 4 3 2 1

To Gisela

Contents

Preface

The development of synthetic-aperture radars was motivated by the military's need to conduct radar reconnaissance at long ranges with fine resolution and from high-performance aircraft.

High-performance aircraft dictated small antennas. In conventional radars, small antennas result in coarse cross-range (azimuth) resolution at long ranges. The desired long ranges were those approaching the radar horizon range.

The need for fine azimuth resolution drove the development of the synthetic-aperture radar (SAR). In an SAR, a long antenna array is synthesized by moving a small physical antenna along a flight path, while transmitting, receiving, and storing the return signals at specific locations of the flightpath. After the collection of sufficient returns, the signals are added to form an azimuth beam which is governed by the length of the synthesized array, and not the length of the physical antenna.

The azimuth resolution attained with this technique at long ranges far exceeded the range resolution at the same ranges obtained using conventional means of transmitting short radar pulses. The peak RF power levels for airborne applications were limited, resulting in low radar returns. Acceptable signal-to-noise or clutter-to-noise levels for desired targets at long ranges could not be obtained with this method.

To obtain radar images with symmetrical resolution in range and azimuth, pulse-compression techniques, such as chirp, were perfected. The devices used for pulse expansion and compression were redesigned for airborne applications in high-performance aircraft.

The development of SAR evolved from having modest resolution and range to operational systems with fine resolution at long ranges and research tools with ultra-fine resolution. The AN/APQ-102A,[1] which was fielded in the mid-1960s, is believed to be the first synthetic-aperture radar to become operational in military use. The system, long declassified, used optical azimuth-signal processing and dispersive

[1] Manufactured by Goodyear Aerospace Corporation, now Loral Defense Systems, Arizona.

delay-line range processing, and had a resolution of 50 by 50 ft at a maximum range of 30 nautical miles (nmi).

Initially, development efforts on SAR concentrated on basic performance, with ECM/ECCM aspects playing a secondary role. But as the resolution at long ranges improved, SAR came to be recognized as a valuable reconnaissance asset, and the development emphasis began to include obtaining such performance in an ECM environment.

This book addresses the subject of ECM/ECCM and SAR. Basic equations for radar resolution and scatterometric performance are not derived. Equations that are needed to treat the affect of signal processing on the jammer signal are derived.

Electronic warfare is a continually changing field. Solutions to new EW problems will have to be generated rather than found in textbooks. This book emphasizes the processes involved in arriving at solutions, rather than providing specific solutions. This structure will aid engineers in developing solutions to their specific EW problem.

Chapter 1 lists the resolution, signal-to-noise and clutter-to-noise ratios for conventional radars. The affect of coherent signal processing on range and azimuth resolution, the signal-to-noise, and clutter-to-noise ratio are derived. Chapter 2 includes EW definitions, identifies possible ECM actions, and identifies ECCM options. Chapter 3 provides a model for a numerical measure for ECM vulnerability, and ECCM effectiveness.

The material in this book has been published in the open literature, and derived from unclassified reports.

Introduction

Imaging radars generate radar maps, which are two-dimensional records of the radar reflectivity of objects on the ground. One of the dimensions is normal to the velocity vector of the radar platform, which is also called the *swath,* and the other is parallel to the velocity vector. The swath of the radar is limited by radar parameters, but the along-track extent of a strip-mapping radar can be made arbitrarily long. The radar map is composed of individual pixels, where the dimensions of each pixel is a scaled-down version of the resolution element size on the ground. The radar reflectivity of a resolution element is represented by the intensity of the pixel on the radar map.

Radar maps are very similar to aerial photographs in appearance, except that whereas the photograph presents an image of the ground just as it would appear to the human eye, radar maps need to be interpreted by experienced analysts to extract the desired reconnaissance or surveillance information. The interpreters use scatterometric (intensity of pixel) and geometric (pattern of the radar return, along-track and cross-track location) data.

There are two categories of radar images: point targets and extended features. Point targets are typically manmade objects and are characterized by having vertical extent and relatively smooth surfaces (compared to the wavelength of the radar signal). Such surfaces can form corner reflectors with other surfaces or with the ground in front of the object and then reflect the incident radar energy specularly. This results in a strong radar return or a high intensity on the radar image. If the radar resolution cell can be made a fraction of the size of the object being imaged, then the radar image displays a pattern of pixels with high intensity, and interpreters are then able to identify the type of the object. Buildings, bridges, large parked aircraft are imaged with a number of pixels, and shapes can be recognized. Smaller manmade objects, such as vehicles, are imaged with only a few pixels, and it is more difficult to differentiate between types of vehicles. The radar cross section is the measure of the reflectivity of point targets; the signal-to-noise ratio is the measure of the radar image quality of point targets.

Extended features are typically of natural origin and are characterized by having horizontal extent and varying degrees of roughness. Such features reflect diffusely; the radar backscatter coefficient is the measure of the reflectivity of the different extended features—such as desert, grass, agricultural fields—and this backscatter coefficient is a function of the grazing angle of the incident radar energy. At shallow grazing angles, less energy is reflected back to the radar than at steep angles. The clutter-to-noise ratio is the measure of a radar's ability to image extended features. Usually the term *clutter* is associated with undesirable radar returns that compete with desired returns. With respect to imaging radars, however, the term *clutter* is used interchangeably with *extended features*. One measure of the quality of a radar image is the ability to display no-return areas, such as concrete or asphalt runways, in low-return areas, typically desert or grass surfaces.

A jammer that is operated against an imaging radar attempts to degrade the image quality by inserting noise into the radar image. This can result in obscuring only areas associated with low backscatter coefficients or in obscuring strong point targets.

Geometric performance can be degraded by introducing amplitude and phase errors, which tend to widen the resolution element size and increase the sidelobe level of the target response. Furthermore, the accuracy with which targets are displayed on the radar map relative to the true location of the targets on the ground—called the *geometric fidelity of the radar system*—can be degraded.

Noise jammers degrade the scatterometric performance of the radar only. Repeater jammers can generate false targets, and the location of such false targets can be controlled in the cross-track direction by the time delay between reception and retransmission of the radar signal. They can also be controlled to some extent in the along-track direction by changing the phase of the returned signal from pulse-to-pulse—that is, by introducing a doppler frequency error.

Jammers do not deliberately introduce phase and amplitude errors to degrade the resolution of the radar return. However, some ECCM devices designed to reduce the effectiveness of jamming do.

Chapter 1
Conventional, Pulse-Compression, and Synthetic-Aperture Radar

1.1 EQUATIONS FOR CONVENTIONAL IMAGING RADAR SYSTEMS

The resolution and received signal, clutter, and thermal noise powers of a radar system using short pulses rather than range compression, and real-aperture rather than synthetic-aperture processing, are defined in the following manner.

1.1.1 Resolution

A resolution element is defined as a spatial and velocity region contributing echo energy that can be separated from that of adjacent regions by actions of the antenna or the receiving system [1]. In conventional radar its dimensions are given by the beamwidth of the antenna, the transmitted pulsewidth and the receiver bandwidth. In ground imaging radars, resolution is the capability of a radar to display the returns from two point targets, separated by the resolution distance, as two responses. It is not necessary that the images of the two point targets be completely separated on the radar image, but rather that there is a clearly defined decrease in intensity between the images of the two targets. The resolution of interest in an imaging radar is the ground range resolution (usually normal to the platform velocity vector) and the azimuth resolution (usually parallel to the velocity vector).

Ground Range Resolution

Targets can be resolved in range when the return of the trailing edge of the transmitted pulse from the closer target is received, before the return of the leading edge of the transmitted pulse from the farther target arrives at the radar antenna. If the targets were positioned along the line of sight of the radar, pulsewidth would be the only parameter to determine the (slant) range resolution. For targets located on the ground, the targets have to be separated further (by a factor of $1/\cos \theta_g$) to be resolved. The ground range resolution is defined as

$$W_r = \frac{c\tau_t}{2 \cos \theta_g} \tag{1.1}$$

where

c = speed of light
τ_t = transmitted pulse width
θ_g = grazing angle

Azimuth Resolution

Targets can be resolved in cross range when the return from a target (which is illuminated by the trailing edge of the azimuth antenna beam) is received before the return from a target (which is illuminated by the leading edge of the azimuth antenna beam) is received. The leading and trailing edges of the radar antenna are usually defined as those angles where the gain of the antenna is 3 dB below the peak gain. The cross-range resolution of a conventional (real-aperture) radar is defined as

$$W_a = R\beta_{az} \tag{1.2}$$

and can be approximated by

$$W_a = R\left(\frac{\lambda}{L}\right) \tag{1.3}$$

based on a nominal beamwidth of $\beta_{az} = \frac{\lambda}{L}$ where

R = slant range to target
β_{az} = 3-dB beamwidth of radar antenna
λ = wavelength of the RF signal
L = length of the physical antenna

1.1.2 Signal, Noise, and Clutter [2]

Signal from a Point Target

The signal returned from a point target is given by

$$S_1 = \frac{P_t G_t}{4\pi R^2} \sigma \frac{A_r}{4\pi R^2} \tag{1.4}$$

where

P_t = peak radar transmitted power
G_t = gain of radar transmit antenna
σ = radar cross section of point target
A_r = effective area of the radar receiving antenna

with $A_r = G_t\lambda^2/4\pi$ for systems using equal transmitting and receiving antenna gains.

The first term in the equation defines the radar RF power density at the location of the target (in RF power per unit area). When multiplied by the radar cross section, the RF power reradiated in the direction of the radar is defined. The signal power density of the reradiated signal, intercepted by the effective area of the radar antenna, is shown in the last term of the equation.

The radar cross section of a point target has the dimension of an area (square feet or square meters) and, except for the case of an isotopic scatterer (such as a conducting sphere), does not represent the projected physical area of the target. Highly directive reflectors, such as corner reflectors, have a much larger radar cross section than their physical area. At X-band, the radar cross section of jeeps and cars is on the order of 100 ft^2, or 10 m^2, and trucks perhaps 1000 ft^2, or 100 m^2. Bridges can have radar cross sections of 10,000 ft^2, or 1000 m^2.

Thermal Noise

The thermal noise for ground imaging radars competing with the signal is

$$N = kT_s B_r \tag{1.5}$$

assuming that the radar receiver bandpass is flat over the radar bandwidth of B_r. In this case

k = Boltzmann's constant

T_s = system noise temperature in kelvins, here 290 NF
N = noise figure
B_r = radar receiver bandwidth. This bandwidth will later be described as the *radar-range bandwidth*, to distinguish it from the doppler bandwidth.

Signal from Extended Features (Clutter)

Whereas the radar cross section of a point target is expressed as an area independent of the radar resolution, we find that the radar cross section of extended features is given by the backscatter coefficient of the terrain, σ_o, times the area that contributes to the radar return at any given time. The backscatter coefficient is a normalized measure of radar reflection from a distributed scatterer [1]. It is defined as the monostatic radar cross section per unit surface area, which is dimensionless. At X-band, the backscatter coefficient may range from −30 dB for arid land to −15 dB for heavy vegetation at shallow depression angles, and from −20 dB for arid land to −5 dB for heavy vegetation at steep depression angles. The backscatter coefficient of smooth surfaces, such as calm water, asphalt, or concrete may be 10 dB below that of arid land.

The area that contributes to the radar return is the product of the ground range resolvable element W_r and the azimuth resolvable element W_a.

It is customary to refer to the radar return from extended features as *clutter*. The clutter level is then

$$C_1 = \frac{P_t G_t}{4\pi R^2} \sigma_o \frac{c\tau_t}{2\cos\theta_g} \frac{R\lambda}{L} \frac{A_r}{4\pi R^2} \tag{1.6}$$

Signal-to-Noise and Clutter-to-Noise Ratios

The absolute value of the signal, clutter, or noise power is not of interest, but the signal-to-noise and clutter-to-noise ratios are. Combining some terms and forming the signal-to-noise ratio for a single pulse results in

$$(S/N)_1 = \frac{P_t G_t^2 \lambda^2 \sigma}{(4\pi)^3 R^4 k T_s B_r} \tag{1.7}$$

and the clutter-to-noise ratio from a single pulse results in

$$(C/N)_1 = \frac{P_t G_t^2 \lambda^3 \sigma_o c\tau_t}{(4\pi R)^3 k T_s B_r 2\cos\theta_g L} \tag{1.8}$$

In addition, because the radar bandwidth for conventional radars is approximately equal to the reciprocal of the transmitted pulse width ($B_r \sim 1/\tau_t$), the signal-to-noise and clutter-to-noise equations can also be expressed in terms of the transmitted RF energy $P_t \times \tau_t$:

$$(S/N)_1 = \frac{P_t \tau_t G_t^2 \lambda^2 \sigma}{(4\pi)^3 R^4 k T_s} \tag{1.9}$$

and

$$(C/N)_1 = \frac{P_t \tau_t G_t^2 \lambda^3 \sigma_o c \tau_t}{(4\pi R)^3 k T_s \, 2 \cos\theta_g L} \tag{1.10}$$

1.2 PULSE-COMPRESSION AND SYNTHETIC-APERTURE PROCESSING

This book is intended to guide the reader through the processes involved in evaluating the performance of a synthetic-aperture radar in an electronic warfare environment. Emphasis is placed on the description of physical processes rather than strictly mathematical derivations, and only those equations that are needed to accomplish this task are derived (resolution, signal, noise, and processing gains). Accordingly, the chapter on radar cannot be viewed as a SAR design guide, because many other aspects of synthetic-aperture radar are not addressed.

The section on range processing treats two methods: dispersive delay line, which acts upon the signal power, and the vector summation of analog or digital signals, which act on signal amplitude. These two methods result in different processing gains for the signal, clutter, and noise. To avoid confusion, the processing gains will be expressed in terms of the radar parameters, and symbols will not be assigned. Compression ratios are independent of the processing method used, and symbols are assigned. Furthermore, even though eventually the signal-to-noise and clutter-to-noise ratios are shown to be independent of the method used, the author is of the opinion that it is important to identify the differences that each process has on the signal and noise power levels.

The section on azimuth processing includes a brief introduction of the synthetic-aperture principle and is used to derive the needed equations.

The equations are given in a format that is thought to be most suitable for subsequent EW performance evaluations.

The equations in the preceding section also describe the resolution and signal, clutter, and thermal noise levels of a synthetic-aperture radar for a single pulse, prior to any signal processing.

1.2.1 Pulse Compression

Radars employ pulse compression to achieve a combination of fine range resolution and adequate signal-to-noise ratio at long range, with available RF peak powers. As seen from equation (1.1), fine range resolution requires short pulses; on the other hand, the signal-to-noise and clutter-to-noise ratios are proportional to the transmitted radar RF energy, which is the product of the radar peak power and the pulsewidth (equations (1.9) and (1.10)). Because the peak power is limited, desirable signal-to-noise ratios at long ranges can be achieved by employing long radar pulses.

Range resolution was stated in terms of pulse width (equation (1.1)), but the pulsewidth is related to the instantaneous radar bandwidth through the Fourier transform. That bandwidth was stated to be approximately equal to the reciprocal of the transmitted pulsewidth for conventional radars but can be made much greater than that for radars using pulse compression. Schemes have been devised to modulate a long pulse (for long radar range) with the bandwidth required for the desired fine resolution, to resolve the conflicting requirements on pulse width. The signal processing in range then amounts to constructing the shortest pulse in the time domain, τ_c, that can be achieved with the transmitted radar bandwidth, B_r.

Linear FM, or Chirp

The method that was initially used for pulse compression, and still is commonly used, is linear FM, or *chirp* [3]. The frequency within the transmitted pulse is linearly swept over the radar bandwidth that is required for the desired resolution.

Linear FM signals can be generated by passive dispersive delay lines as well as by active means, such as sweeping an oscillator over the required radar bandwidth during the transmit pulsewidth, or digitally by generating the quadratic phase as function of time, thereby forming the linear FM waveform.

The specific method used to generate a linear FM waveform will depend on the time-bandwidth product and the need to generate different waveforms, such as in a multimode system.

Passive devices, such as surface acoustic wave (SAW) dispersive delay lines, are capable of only one waveform per device, but they offer the cheapest method of on-board range compression. This is because another device with the opposite frequency-versus-time slope (or the same device, processing the opposite sideband of the transmitted RF signal) can process the entire range line, independent of the number of range elements in the line. SAW dispersive delay lines have an upper limit for the bandwidth and the dispersion time, and the upper limit of the of the time-bandwidth product may be on the order of 10,000.

Actively generated linear FM waveforms (analog or digital), on the other hand, can be programmed to generate a multitude of waveforms, but they require additional hardware to compress the received radar returns. The complexity of such processors increases with the compression ratio and the number of range elements. Actively generated waveforms are capable of a wider bandwidth, longer dispersion times, and larger time-bandwidth products than those generated by passive devices.

Pulse compression by dispersive delay lines is accomplished by passing the received signal through a delay line having the opposite FM slope of the transmitted signal, thereby recreating the instantaneous bandwidth. But pulse compression can also be achieved by processing analog signals (autocorrelation of the transmitted signal or cross correlation, when weighting is used). Optical signal processing is an example of the Fourier transform of analog signals.

The fast Fourier transform (FFT) is an efficient implementation of digital signal processing. Prior to processing, the radar return signal has to be digitized, and the rate is twice the highest frequency of $B_r/2$. The number of samples that need to be processed for each resolvable element is then the sampling rate times the transmitted pulse width:

$$n_r = 2\left(\frac{B_r}{2}\right)\tau_t = B_r\tau_t \tag{1.11}$$

When linear FM signals are compressed, the resultant pulsewidth is

$$\tau_c = \frac{1}{B_r} \tag{1.12}$$

and the ground range resolvable element w_r is

$$w_r = \frac{c\tau_c}{2\cos\theta_g} = \frac{c}{2\,B_r\cos\theta_g} \tag{1.13}$$

Actually, the compressed pulse resulting from a uniformly weighted radar bandwidth is of $(\sin x)/x$ form, and the 3-dB width of that pulse is $0.886/B_r$. When weighting is applied to the bandwidth, the resultant pulse widens and can be wider than $1/B_r$. In the following we will use the approximation of equation (1.12).

Pulse compression improves the range resolution by a factor that can be expressed as the time-bandwidth product of the transmitted pulse, or the ratio of the transmitted to compressed pulsewidth.

Discrete Coded Waveforms

The other method that is frequently used to obtain pulse compression is discrete coding [4]. In this case, a long transmit pulse, τ_t, is constructed from a series of n_r

short pulse segments of duration τ_c, and the coding is usually accomplished by changing the phase of the carrier frequency from one segment to another. The short pulses are of the duration that is required to achieve the desired range resolution; the radar bandwidth associated with those short pulses is B_r.

Processing consists of passing the return signal through a delay line of duration $\tau_t = n_r \times \tau_c$, which has taps at intervals of τ_c. The signals from these taps are summed, after the phase at each tap has been adjusted so as to match the code of the transmitted pulse. When the return signal is fully contained within the delay line, all signals are in phase, and their amplitudes add linearly. The delay line output for this condition is the mainlobe peak of the compressed pulse. The codes have to be selected such that the sum of the signals is very small when the pulse code does not match the delay line code; that is to say, that the sidelobes of the compressed pulse are low.

The result is a compressed pulse of duration τ_c, and the ground range resolution is again

$$w_r = \frac{c\tau_c}{2 \cos \theta_g}$$

as in equation (1.13), but now can also be expressed as

$$w_r = \frac{W_r}{n_r} \tag{1.14}$$

1.2.2 SAR Processing

Radars with small physical antennas employ SAR processing to achieve fine azimuth or cross-range resolution at long ranges. Such processing is achieved by summing *n* radar returns—*n* being the number of synthetic-aperture elements along the flight path—after the phase of each signal has been corrected for the phase shift caused by the pathlength differences of each return from the pathlength of the radar return in the center of the synthetic array (focused system).

Figure 1.1 compares a physical antenna with the synthetic antenna. As shown in Figure 1.1(a), all radiating elements of the physical antenna radiate signal power simultaneously. The phase of all such signal sources at some reflector (target) location is the composite of all signals, and the signal will appear to have emanated from the phase center of the antenna. The signal reflected from that target will have this composite phase. However, depending on the direction of arrival at the face of the antenna, the signals intercepted by the individual radiating elements may differ in phase. The antenna feed then forms the vector sum (phase and amplitude) of the signals from all radiating elements, in real time. The signal out of the feed will be a maximum when all signals are in phase and will be zero when the pathlength difference over the length of the antenna is one wavelength.

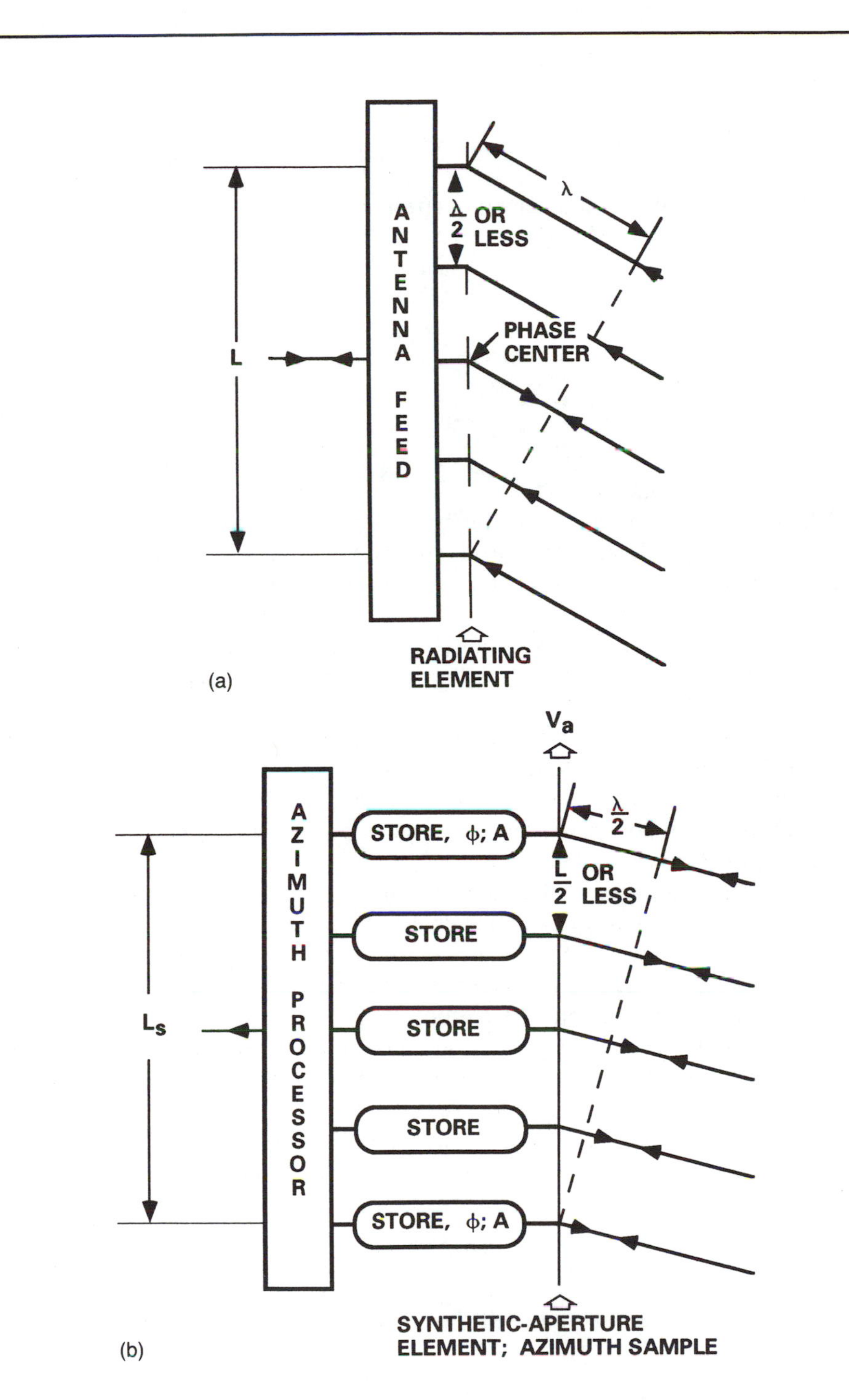

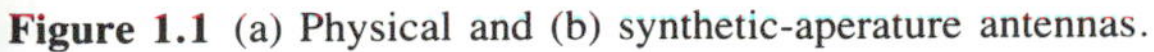
Figure 1.1 (a) Physical and (b) synthetic-aperature antennas.

$$\beta_{\text{null}} = \pm\frac{\lambda}{L} \tag{1.15}$$

The 3-dB angular width of the antenna pattern can be approximated by

$$\beta_{az} \sim \frac{\lambda}{L} \tag{1.16}$$

and the angular resolution at some range R becomes

$$W_a = R\,\frac{\lambda}{L}$$

This was stated in equation (1.3).

The synthetic aperture (Figure 1.1(b)) is formed by transmitting and receiving radar signals along a flight path and by storing the radar return by phase and amplitude. Each such event (transmit, receive, and store) results in an element of the synthetic aperture, which we can also call an *azimuth sample*. When the radar has traversed the length of the synthetic aperture, sufficient samples have been stored, and the azimuth processor then forms the vector sum of the sequentially collected samples.

Measuring the phase of the radar return signal is equivalent to measuring the round-trip distance from the radar antenna to the target to within a fraction of a wavelength of the transmitted signal. It is the change of distance from pulse to pulse that is of interest. The nominal distance, measured by the time of arrival (slant range), is not of interest here.

To measure the phase of the radar return signal to within a few degrees, it is necessary to derive the transmitted RF signal from a stable reference and to compare the phase of the returned signal to that of the reference; that is, there has to be coherence between the transmitted signal and the reference, and this coherence has to be maintained for each interpulse period.

To measure the change of phase from pulse to pulse requires that the reference frequency remain constant, and that condition has to be maintained—as a minimum—over one synthetic-aperture length.

Because each subsequent sample experiences a pathlength change on transmit and on receive, we find that the synthetic aperture has a null in azimuth at an angle

$$(\beta_{\text{null}})_s = \pm\frac{\lambda}{2L_s} \tag{1.17}$$

and we can now approximate the 3-dB beamwidth of the synthetic antenna by

$$(\beta_{az})_s = \frac{\lambda}{2L_s} \tag{1.18}$$

The maximum synthetic-aperture length for a continuous strip-mapping radar (not spotlight) is that flight path over which a target remains illuminated by the 3-dB one-way azimuth pattern of the physical antenna, or

$$L_{s\max} = R\beta_{az} \sim R\frac{\lambda}{L} \tag{1.19}$$

If all such signals are processed, the best obtainable resolution results [5].

$$(w_a)_b = R(\beta_{az})_s = \frac{R\lambda L}{2R\lambda} = \frac{L}{2} \tag{1.20}$$

Errors in motion compensation prevent this maximum synthetic-aperture length from being consistently illuminated, and in practice a shorter length than the maximum will be processed, which results in a cross-range resolution that is coarser than the best obtainable. The synthetic-aperture length for some coarser resolution $w_a > L/2$ can be found by substituting the desired resolution for the best resolution in equation (1.19), which then becomes

$$L_s = \frac{(R\lambda)}{2w_a} \tag{1.21}$$

A differentiation has to be made between the number of samples (elements in the synthetic aperture) collected and the number of samples processed.

Number of Samples Collected

The rate of collection has to be high enough to avoid aliasing of doppler energy, and according to the Nyquist sampling criterion the sampling frequency has to be a minimum of twice the highest frequency to be sampled.

Figure 1.2 shows the doppler frequency spectrum before sampling. This spectrum is governed by the azimuth antenna pattern, in which the intensity is proportional to the two-way gain, and the doppler frequency is proportional to the sine of the angle of the antenna pattern. Because the sidelobe pattern can extend to ±90° from the beam peak, the maximum illuminated doppler frequency is very high indeed. However, the energy associated with such high doppler frequencies is very low, and it is customary to consider the maximum band of doppler frequencies of interest ($B_{d\max}$) to include those frequencies associated with surface scatterers within the 3-dB angle of the one-way antenna pattern. The maximum doppler frequency is then

$$(f_{d\max}) = \frac{2v_a}{\lambda} \times \frac{\lambda}{2L} = \frac{v_a}{L} \tag{1.22}$$

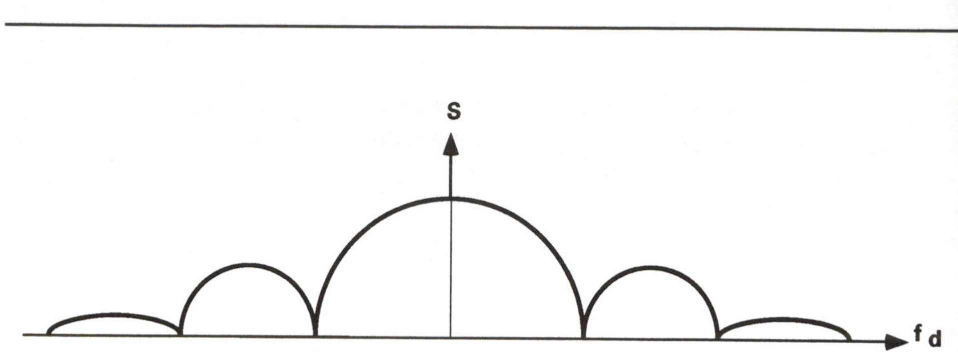

Figure 1.2 Continuous radar signal doppler frequency spectrum.

and the minimum sampling rate, or minimum pulse repetition frequency (PRF) (for purposes of data collection), is then

$$\mathrm{PRF}_{\min} = \frac{2v_a}{L} \tag{1.23}$$

where v_a = vehicle velocity.

It is found that the radar has to be pulsed at least once every time the aircraft flies half the length of the physical antenna to avoid ambiguous azimuth lobes. This is similar to the requirement that the element spacing of physical antennas not exceed one half wavelength, to avoid grating lobes (see Figure 1.1).

When sampled at that rate, some of the main beam energy and all the sidelobe energy is folded into the region of possible doppler frequencies of $\pm(\mathrm{PRF}_{\min})/2$, which is also the maximum doppler bandwidth of the radar signal ($B_{d\max}$). This is illustrated in Figure 1.3.

The number of synthetic-aperture elements collected, n_c, is given by the time required to fly one synthetic aperture, L_s, times the PRF:

$$n_c = t_s \times \mathrm{PRF} \tag{1.24}$$

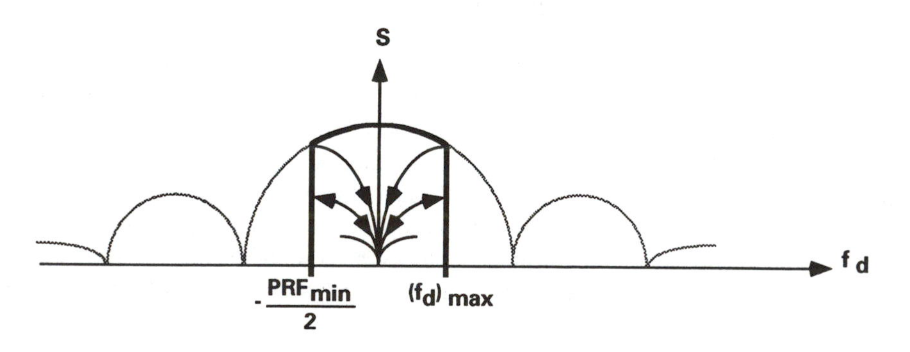

Figure 1.3 Sampled radar signal doppler frequency spectrum.

where the synthetic aperture time is

$$t_s = \frac{L_s}{v_a} \tag{1.25}$$

For the purpose of further discussions we ignore the aliased doppler signals and represent the radar signal as extending from $\pm(\text{PRF}_{\text{min}})/2$, and 0 outside that region, as shown in Figure 1.4.

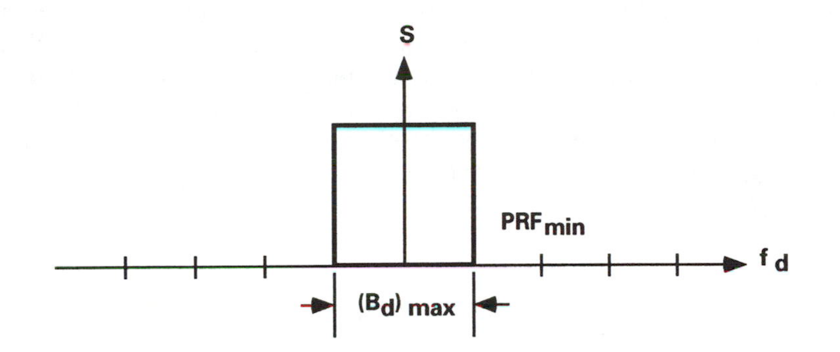

Figure 1.4 Equivalent radar signal doppler frequency spectrum.

Thermal noise is noncoherent from pulse to pulse, which means that the phase of the sampled noise will vary randomly, and therefore the total noise power will be uniformly distributed over the range of possible doppler frequencies, as shown in Figure 1.5. In this case (minimum PRF), that range equals the maximum bandwidth of the radar signal; that is, the noise spectrum extends over the same doppler frequency bandwidths as the radar signal.

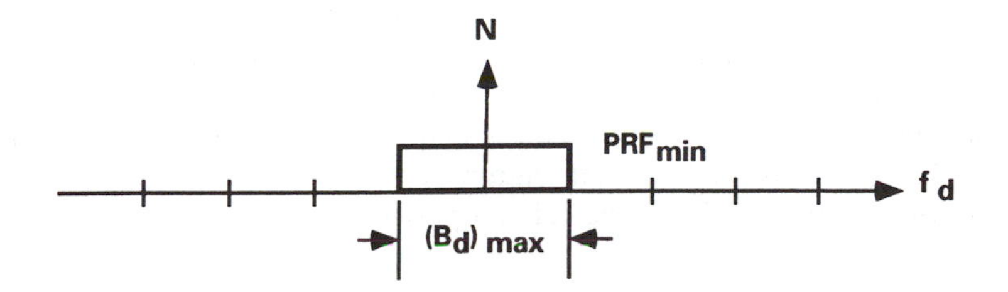

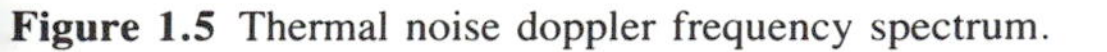

Figure 1.5 Thermal noise doppler frequency spectrum.

Number of Samples Processed

When the desired cross-range resolution is $w_a > L/2$ then a doppler bandwidth of only

$$B_d = \frac{v_a}{w_a} \tag{1.26}$$

needs to be processed (which is one sample every time the platform moves by one resolvable element), and the number of samples processed, as a minimum, becomes that doppler bandwidth times the synthetic-aperture time. To reduce the number of samples to that minimum requires, however, that the radar return signals be filtered in the doppler frequency domain to $\pm B_d/2$ and that the remaining data be resampled at a rate of B_d samples per second. This reduces the minimum number of samples to be processed by a factor of $(2w_a/\mathrm{L})^2$.

$$n_{\min} = B_d \times t_s = \frac{(R\lambda)}{2w_a^2} \tag{1.27}$$

We find that the minimum number of samples to be processed increases linearly with the wavelength and the range, and inversely with the square of the cross-range resolution element size. The actual number of samples processed can be any number between the minimum required (equation (1.27)) and the maximum collected (equation (1.24)), and in digital processors may be selected to be a power of 2 at the maximum range.

Processing a larger number than the minimum required to achieve the desired resolution will not improve the resolution, and equation (1.27) also defines the cross-range compression ratio, γ_a.

The signal power to be processed is reduced by such filtering, and because the noise power extends over the same doppler frequency range as the radar signal, the noise power is reduced by the same ratio (Figures 1.6 and 1.7). Processing for coarser resolution reduces the actual levels of the signal and the noise, but the signal-to-noise ratio at the input of the processor remains unchanged. Although processing additional samples does not improve the resolution, it does enhance the signal-to-noise and clutter-to-noise ratios at the output of the processor, because these ratios are proportional to the actual number of samples processed.

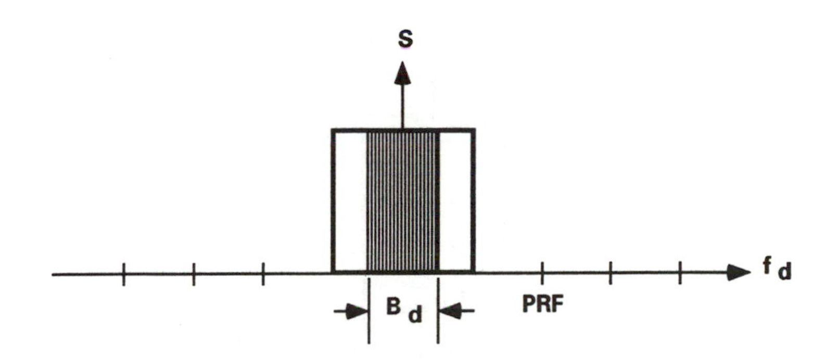

Figure 1.6 Radar signal doppler processing for coarser resolution.

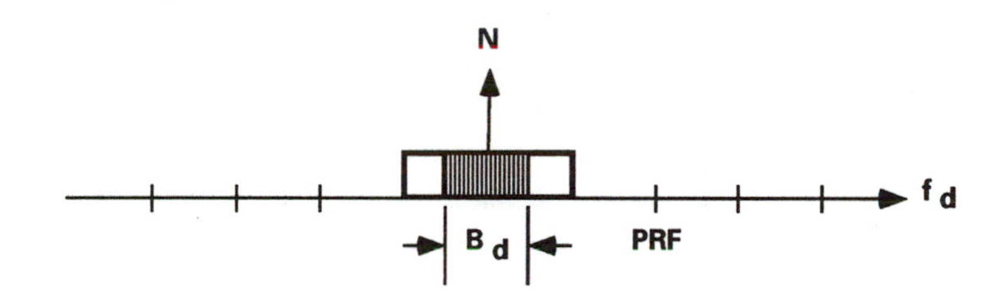

Figure 1.7 Thermal noise doppler processing for coarser resolution.

Oversampling

When processing for coarser resolution, the actual (data-collecting) PRF need not be the minimum given in equation (1.23), and in practice the PRF may exceed that minimum by a factor K_a, which is called the *oversampling factor*.

$$\text{PRF} = \text{PRF}_{\text{min}} \times K_a$$

or

$$K_a = \text{PRF} \times \left(\frac{L}{2v_a}\right) \tag{1.28}$$

Such oversampling does not alter the radar signal doppler frequency spectrum (except for the aliased doppler signals, which are ignored here), because the doppler bandwidth of the radar signal is governed by the illumination, not the sampling rate. However, now the noise covers the actual PRF range, which is K_a times that of the radar signal (Figures 1.8 and 1.9).

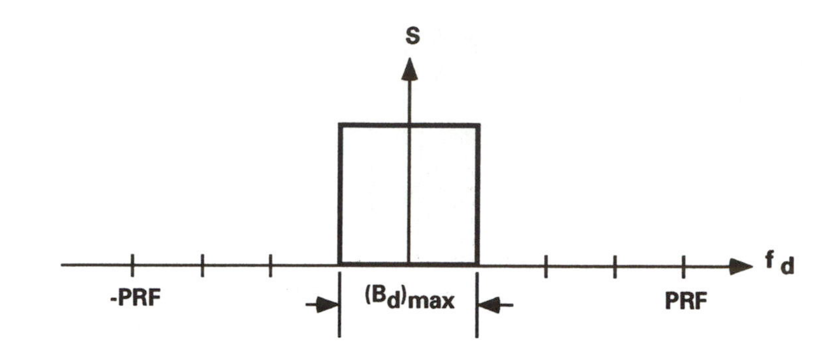

Figure 1.8 Radar signal spectrum with oversampling.

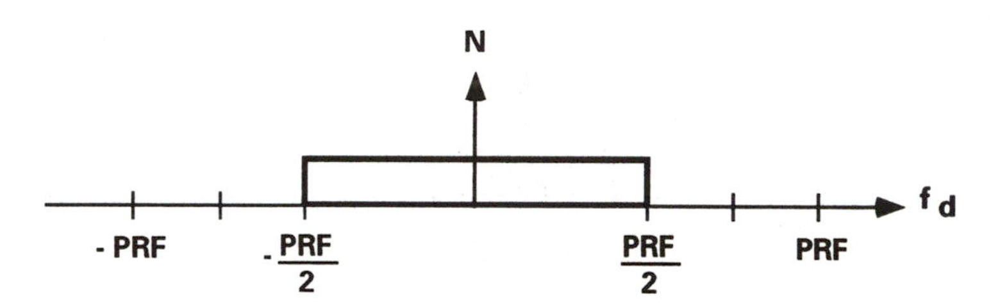

Figure 1.9 Noise spectrum with oversampling.

Deramp-on-Receive [5]

Unlike dispersive delay line processing and correlation processing where the radar range bandwidth remains unchanged by demodulation from RF (against a constant reference), deramp-on-receive, or *stretch processing,* results in an output signal bandwidth that is different from the input bandwidth. One application of such processing is to reduce the radar return signal bandwidth prior to digitization, because the deramped bandwidth is determined by the processed range swath, not the transmitted radar bandwidth. Figure 2.4(a) shows the time line of a radar using the deramp technique. A transmit pulse is swept over a bandwidth B_r during the transmit pulsewidth τ_t. The radar return, shown here as three point targets, is demodulated against a reference that has the same slope as the transmit pulse, but a duration of at least the transmitted pulsewidth plus the swath time.

1.2.3 Signal, Clutter, and Noise

To derive the signal, clutter, and noise levels after processing, we start out with the levels from one pulse, as given in the section on conventional radars. After range processing a subscript r is added and after azimuth processing a subscript a is added to the particular symbol. The equations of interest are the signal-to-noise ratio and eventually the signal-to-jam ratio. However, because range and azimuth processing has different effects on signals (which are coherent in range and in azimuth) than it has on thermal or jammer noise, it is necessary to treat the signal and the noise separately and to form the signal-to-noise ratio and signal-to-jam ratio at the output of the processor. Figure 1.10 is a simplified block diagram, which also identifies the symbols used in this model.

Both range and azimuth signal processing are based on the Fourier transform principle, which relates spectra in the frequency domain to responses in the time domain. But the processing gain for signals and for noise resulting from such transforms is dependent upon the particular mechanization of the transform.

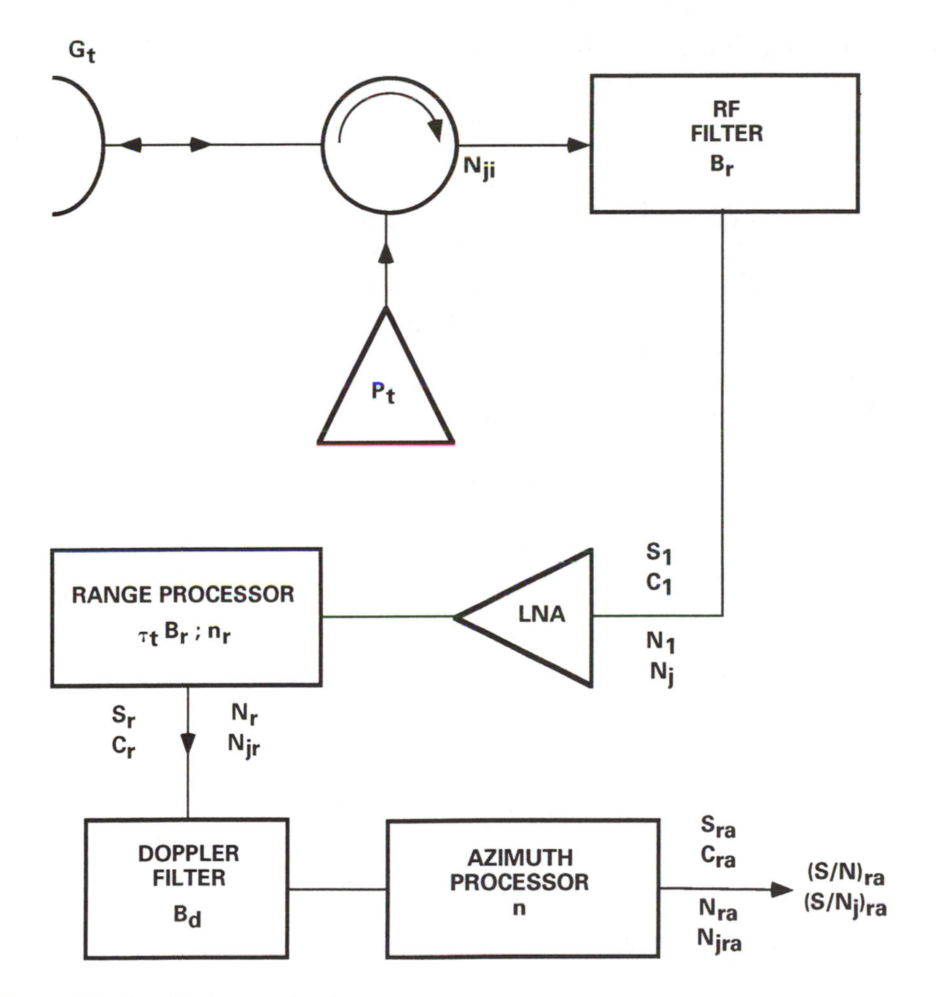

Figure 1.10 Simplified processing diagram.

Processing of Point Target Signals

The radar return from a point target, S_1, is given by equation (1.4). Range processing results in the compression of the transmitted pulsewidth to the processed pulsewidth.

When a linear FM waveform is processed by dispersive delay lines, the signal energy remains constant before and after range processing, since the dispersive delay line merely orders the signal power associated with different frequencies in the time domain. Therefore the signal energy, which is the product of the signal power times the pulsewidth, remains constant:

$$S_1 \tau_t = S_r \tau_c \tag{1.29}$$

and the signal after range processing becomes

$$S_r = S_1 \left(\frac{\tau_t}{\tau_c} \right) = S_1 \tau_t B_r \tag{1.30}$$

Autocorrelations, cross-correlations, and FFTs are implementations of the Fourier transform that act on signal amplitude. These processes can be thought of as the vector addition of a number of signals, and when all signals are in phase (which results in the peak processing gain), the sum-signal amplitude increases by the number of signals added (for uniformly weighted input signals), and the signal power by the number of signals squared.

The signal power level after such range processing becomes

$$S_r = S_1 n_r^2 = S_1 (\tau_t B_r)^2 \tag{1.31}$$

and after azimuth compression becomes

$$S_a = S_1 n^2 \tag{1.32}$$

The signal power after range and azimuth processing becomes

$$S_{ra} = S_1 (\tau_t B_r) n^2 \tag{1.33}$$

for linear FM processed with dispersive delay line, and

$$S_{ra} = S_1 (\tau_t B_r)^2 n^2 \tag{1.34}$$

for processing by correlation or FFT.

Processing of Thermal Noise

The dispersive delay line acts on thermal noise in the same manner as on radar signals; that is, the delay line orders the noise power associated with different frequencies in the time domain. If a sample of thermal noise of the duration of one compressed pulsewidth were processed, then that sample would be spread (dispersed) over the duration of the transmitted pulsewidth, and the actual noise level in each resolution cell would be reduced by the compression ratio from the level at the input to the delay line. However, noise is continuous in time, so what was said about that one sample holds true for each subsequent sample of noise, and the noise power from all such overlapping samples add linearly. Eventually, the thermal noise level at the output of the dispersive delay line equals the level at the input:

$$N_r = N_1 \tag{1.35}$$

Figure 1.11 illustrates this process.

However, the noise level after processing by the summation of a number of elements increases linearly with the number of elements. Thermal noise is noncoherent from element to element within a transmitted coded pulse, or from sample to sample in a linear FM pulse, and the range processed thermal noise becomes

$$N_r = N_1 \times n_r = N_1(\tau_t B_r) \tag{1.36}$$

Thermal noise is also noncoherent from azimuth sample to azimuth sample, and the

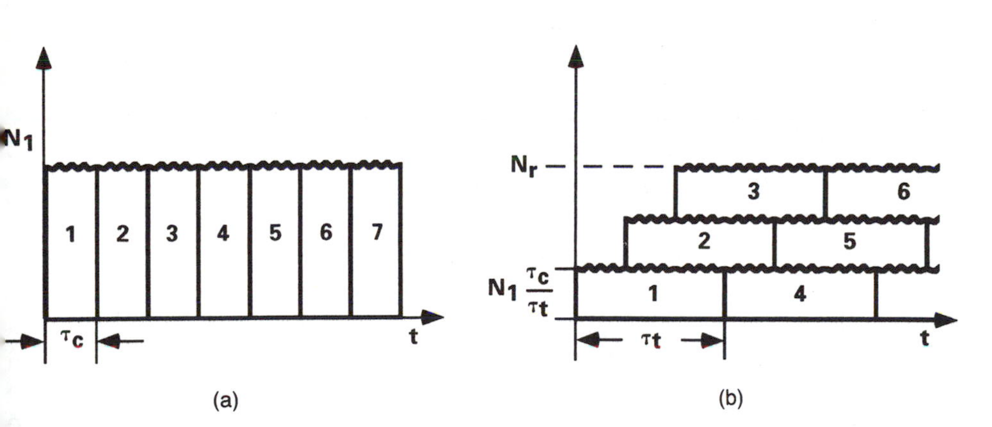

Figure 1.11 Dispersive delay line processing of thermal noise. (a) Dispersive delay line input. (b) Dispersive delay line output.

noise level after azimuth processing also increases linearly with the number of samples added. After azimuth processing, the processed noise power becomes

$$N_a = N_1 \times n \text{ after azimuth processing} \tag{1.37}$$

Forming the signal-to-noise ratio after range and azimuth processing (equations (1.33), (1.35), and (1.37), or (1.34), (1.36), and (1.37)) yields

$$(S/N)_{ra} = (S/N)_1 \tau_t B_r n = (S/N)_1 \gamma_r \gamma_a \frac{2w_a}{L} K_a \tag{1.38}$$

with $\gamma_r = \tau_t B_r$ the range compression ratio and $\gamma_a = n_{\min}$. The factor $(2w_a/L)$ accounts for oversampling of the processed doppler bandwidth relative to the maximum illuminated bandwidth, and $K_a$the oversampling of the minimum PRF. This number, n, is the maximum number of samples that can be processed and is to be used when the doppler signals are not resampled at a lower rate. Resampling can reduce that number to any value between the above and the minimum required, $n_{\min}$.

We find that the signal-to-noise ratio is enhanced by the range compression ratio, by the azimuth compression ratio and by oversampling, and the degree of enhancement is independent of the type of pulse coding or processing method used.

We can now express the signal-to-noise ratio in terms of the average transmitted RF power, either starting with equation (1.7) or using equation (1.9) and noting that with pulse compression, B_r is approximately $1/\tau_c$, not $1/\tau_t$. We can express the processed signal-to-noise ratio as

$$(S/N)_{ra} = \frac{P_t \tau_t \mathrm{PRF}\, G_t^2 \lambda^3 \sigma}{(4\pi R)^3 k T_s 2 w_a v_a} \tag{1.39}$$

It is observed that the desired signal-to-noise ratio depends on the radar's average radar RF power $P_t \tau_t$ PRF, and that value can be obtained by a high peak power, a long transmitted pulse, a high PRF, or any combination thereof.

Processing of Signals from Extended Features

Prior to any processing, the clutter return is contributed by a large area, governed by the transmitted pulsewidth in range and the azimuth beamwidth of the physical antenna in cross range, as shown in equation (1.6). We define the clutter return in terms of the area contributing to the return and lump the remaining quantities into a single constant, Const.:

$$C_1 = \text{Const.} \times W_r \times W_a$$

Clutter is composed of a large number of individual scatterers, and the phases of the returns from these scatterers vary randomly. The clutter power level received from a resolution cell is equal to the noncoherent addition of the returns from all individual point reflectors.

After processing, the area contributing to the clutter power is the processed resolution cell $w_a \times w_r$, and the clutter power is again equal to the noncoherent addition of all scatters in that cell, but the contributions of all point reflectors have been enhanced in the same manner as isolated point targets (see equations (1.33) and (1.34)).

The clutter power level after processing is then

$$C_{ra} = \text{Const. } w_r(\tau_t B_r) w_a n^2$$

for linear FM signals processed by dispersive delay line, and

$$C_{ra} = \text{Const. } w_r(\tau_t B_r)^2 w_a n^2$$

for correlation and FFT processing.

Coherent processing increases the clutter power level by the range and azimuth processing gains; signal processing compresses the resolution element size and enhances the clutter-to-noise ratio.

The range resolution element is reduced by the range compression ratio to $w_r = W_r/\gamma_r$. Doppler filtering limits the synthetic-aperture length to that required for the desired resolution, and that factor is $2w_a/L$. Azimuth signal processing compresses that synthetic-aperture length by the azimuth compression ratio $\gamma_a = n_{\min}$. The resultant azimuth resolution element size is

$$w_a = \frac{W_a}{\dfrac{2w_a}{L} \times \gamma_a}$$

Signal processing increases the clutter-to-noise ratio by the range compression ratio, γ_r, and the number of signals coherently processed, n. The clutter-to-noise ratio after processing is then

$$(C/N)_{ra} = \frac{\text{Const. } w_r \gamma_r w_a \gamma_a}{N_1} = \frac{\text{Const. } W_r W_a}{N_1} K_a \tag{1.40}$$

Substituting results in

$$(C/N)_{ra} = \frac{P_t G_t^2 \lambda^3 \sigma_o}{(4\pi R)^3 k T_s B_r} \frac{c\tau_t}{2\cos\theta_g} \frac{K_a}{L} \tag{1.41}$$

Note that the clutter level is contributed by the processed resolution cell and is enhanced by coherent processing. Although the equations for the clutter-to-noise ratio before (equation (1.8)) and after (equation (1.41)) processing have the same form (except for the oversampling factor K_a), they do represent different conditions.

In terms of the average transmitted radar power, the clutter-to-noise ratio is

$$(C/N)_{ra} = \frac{P_t \tau_t \, \text{PRF} \, G_t^2 \lambda^3 \sigma_o w_r}{(4\pi R)^3 k T_s 2 v_a} \tag{1.42}$$

The equation for the clutter-to-noise ratio can also be obtained by setting the cross section in equation (1.39) equal to $\sigma_o w_r w_a$.

Weighting

Weighting of input data is used to reduce the sidelobe level of the processed signal, compared to the $(\sin x)/x$ response of uniformly weighted data. But weighting also reduces the processing gain, reduces the total power in the aperture, and widens the mainlobe response. Appendix A of Barton et al. [6] quantifies the descriptive parameters of far-field antenna patterns for commonly used antenna aperture illumination functions. Appendix B establishes the analogy between antenna illumination functions and far-field patterns, and frequency spectra and waveform in the time domain.

The sidelobe level and mainlobe response are not of interest when evaluating the effect of weighting on the susceptibility of a radar to jamming, but the signal-to-jam and clutter-to-jam ratios are. The processing gain for a point target is derived from the linear addition of the amplitudes of the processed signals. Barton et al. [6] defines A as the sum signal of the processed signals, and A^2 as the processing gain.

If weighting in the form of $H(f)$ is introduced, the processing gain becomes

$$A_w^2 = \left| \int_{-B_r/2}^{B_r/2} H(f) \, df \right|^2 \tag{1.43}$$

We define A as the sum signal of a uniformly weighted aperture and A_w as the sum signal of the weighted aperture. Weighting then reduces the processing gain for point targets by $(A_w)^2/A^2$. This reduction in processing gain is given in dB for a large number of weighting functions, with A^2 for uniform weighting normalized to 1 [6].

Processing of Weighted Noise

It has been stated (equation (1.36)) that the noise power contributed by all samples add linearly and the noise power out of the processor, for uniformly weighted apertures (processed bandwidth in range and processed synthetic aperture in cross range), is equal to the sum of the noise samples.

With weighting, the output level of thermal noise is still the sum of all samples in the aperture, except that the samples are no longer equal.

$$N = \sum_{i=1}^{n} N_i \tag{1.44}$$

This sum is identified in Barton et al. [6] as C, the total power radiated.

Processing of Weighted Clutter

Clutter was shown to be enhanced by processing, and that enhancement is equal to the processing gain in range and azimuth; it was also observed that the equations for clutter to noise before and after processing are of the same form. Therefore, the processed weighted clutter level is also given by C, the signal equivalent of the total power radiated, just like the weighted processed noise.

Weighting does not change the clutter-to-noise ratio because it reduces the total power in the aperture, and processed clutter—as well as processed noise—is equal to the total power in the aperture.

The reduction in the signal-to-noise ratio is shown to be equal to the aperture efficiency

$$\eta = \frac{A_w^2}{C} \tag{1.45}$$

Because weighting can be applied to range and azimuth signals, the combined effect on the signal-to-noise or signal-to-clutter ratio is the product of the range and azimuth aperture efficiency factors for the range and azimuth weighting functions:

$$L_m = \frac{1}{\eta_r}\frac{1}{\eta_a} \tag{1.46}$$

If, for instance, it is desired that the peak sidelobes of the processed range or cross-range signals be no higher than 30 dB below the peak response, one might employ a 30-dB Taylor weighting. Table 1.1 highlights the effect of such weighting.

If all other performance parameters of the radar remain constant, then the bandwidth of the signal would have to be widened by the factor of 1.115/0.886 = 1.26 to retain the resolution of the unweighted design, and the average power would have to be raised by a factor of 1/0.85 = 1.176 to compensate for the reduction in aperture efficiency in addition to the added power to compensate for the higher thermal noise power due to the widened bandwidth.

Table 1.1
Effect of Weighting

Parameter	*Uniform Weighting*	*30-dB Taylor Weighting*
Processing gain, A_w^2/A^2	1.0	0.407
Power in aperture, C	1.0	0.478
Aperture efficiency, η	1.0	0.85
3-dB response width	0.886 $c/2B_r$	1.115 $c/2B_r$
Highest sidelobe, dB	−13.3	−30.9

Speckle

When we examine the radar image of a homogeneous extended feature (clutter), one that has a constant backscatter coefficient, we find that the radar image of such clutter will show brightness variations from one resolution element to another. That variation is called *speckle* [7]. Clutter is composed of a large number of randomly phased point reflectors, and the magnitude of the vector sum of all such scatterers for different resolution cells follows the Rayleigh distribution. However, the phase of that vector sum of a particular resolution cell, when observed over one synthetic-aperture length, will follow the well-know quadratic function. (Clutter is subject to compression.) "One-look" synthetic-aperture radar images of an area with constant back scatter coefficient will be very grainy; that appearance can be smoothed out by the use of multiple looks [7–9].

Thermal noise is also subject to speckle, because thermal noise within the radar bandwidth consists of a large number of signals of randomly phased frequencies. The magnitude of the vector sum of all such signals also follows the Rayleigh distribution. However, that vector sum is not coherent from pulse to pulse and is therefore not subject to coherent processing gain. Instead, a large number of noise samples are added noncoherently, thereby forming in essence multiple looks of the noise in a resolution cell, which tend to smooth out the intensity variation from resolution element to element. ("Snow" on a TV screen is an example of the "one-look" image of thermal noise.)

Thermal noise would be imaged in no-return areas of a radar image, such as in the shadows of mountains or in water surfaces. However, when such areas occupy only a small portion of the imaged swath and the major part of the imaged swath includes strong returns, then the radar system and the processor will adjust the gains such that the desired radar image features are displayed within the optimum dynamic range of the display. In that case, the noise within the no-return areas will usually be below the threshold of the display.

When the radar images water surfaces over the entire swath, however, the gains will be adjusted such that the weakest signal, which is thermal noise, is displayed within the optimum dynamic range of the display, and the speckle is observable.

Thermal noise at the input of the range and azimuth processor has the same bandwidths as the radar signal, and because the processor performs a Fourier transform of the range and azimuth spectra, speckle size of thermal noise on the order of the processed radar signals in range and cross-range result.

1.3 GEOMETRIC PERFORMANCE OF THE SYNTHETIC-APERTURE RADAR

In addition to the scatterometric data of imaging radars, which are expressed by the signal-to-noise and clutter-to-noise ratios of targets of interest, image interpreters use geometric performance data to extract the needed information from the radar images. One set of such data relates to the processed resolution and is expressed as the 3-dB width, the peak sidelobe level, and the integrated sidelobe ratio. The other set of geometric performance data is the geometric fidelity.

The radar will be designed to meet certain requirements with respect to the geometric performance; phase and amplitude errors degrade the geometric performance of the radar. Widening of the 3-dB width, or poorer resolution, results in the reduction of detail in the radar image. Higher peak sidelobes can cause spurious responses of strong point targets or obscure weak point targets in the immediate vicinity of strong point targets. The integrated sidelobe ratio contributes to the background noise of the radar image, and higher ratios increase that noise, which results in poorer image contrasts.

1.3.1 Resolution and Sidelobes

The numerical example in Table 1.1 lists the 3-dB width and the peak sidelobe level for a uniformly weighted signal and for a Taylor-weighted signal. The uniformly weighted signal has a constant amplitude over the entire bandwidth and a constant FM slope (or a quadratic phase function) across the bandwidth. The amplitude of a Taylor-weighted signal decreases from the center of the signal bandwidth, and the attenuation increases with the distance from the center (resembling a pedestal plus a cosine-squared shape), and the phase is an unmodified quadratic function. Weighting then can be thought of as a deliberately introduced, carefully controlled amplitude error.

Amplitude and Phase Errors [10]

The primary sources of phase and amplitude errors in the range frequency domain are the RF and IF components of the radar, as well as the transmit signal generator and the signal processor. In the doppler frequency domain, the sources of such errors are primarily uncompensated platform (aircraft, spacecraft) motions, and antenna pointing errors.

Assessing the effect of phase and amplitude errors on the compressed signal waveform is not an easy task, because the errors can assume many different forms and magnitudes across the aperture. Some errors are systematic, but some can vary randomly. One approach is to postulate specific error forms and then determine numerically their effect.

Errors can be described by their magnitude and their form across the aperture. The range bandwidth is the aperture in the range direction; the doppler bandwidth, or the synthetic-aperture time, is the aperture in the cross-range direction. Table 1.2 gives an indication of the effect of phase and amplitude errors of one cycle and two cycles per aperture.

The value in the 3-dB width column is normalized to $c/2B_r$ for range resolution and to $\lambda R/2L_s$ for azimuth resolution. The peak sidelobe level is relative to the peak of the main response, and the integrated sidelobe ratio is the total signal power in the sidelobes relative to the total power in the signal being processed. The magnitude of the phase error is given in degrees; the magnitude of the amplitude error, expressed as a decimal in the error column of the table, is the ratio of the error signal amplitude to the peak amplitude of the uncorrupted signal.

It can be seen that an amplitude error of one cycle per aperture affects primarily the 3-dB width and that phase errors and higher frequency amplitude errors degrade primarily the sidelobe characteristics.

Table 1.2
Phase and Amplitude Error Effects on Synthetic-Aperture Radar

Error	*Uniform Weighting*			*−30-dB Taylor Weighting*		
Type, Cycles/Aperture, Form, Magnitude	*3-dB Width*	*Peak Sidelobe Level*	*Integrated Sidelobe Ratio*	*3-dB Width*	*Peak Sidelobe Level*	*Integrated Sidelobe Ratio*
None	0.886	−13.3	−9.86	1.115	−30.9	−21.25
Phase, 1 CPA, cos, 25°	0.9	−10.0	−7.8	1.16	*	−14.1
Phase 2 CPA, cos, 20°	0.886	−9.0	−8.1	1.115	−14.0	−13.2
Amplitude 1 CPA, cos, 0.25	0.96	−12.0	−9.4	1.25	*	−17.5
Amplitude 2 CPA, cos, 0.1	0.89	−13.3	−8.1	1.12	−23.0	−15.6

*The first sidelobe is absorbed in the mainlobe

1.3.2 Geometric Fidelity

The geometric fidelity of a radar system is a measure of the accuracy with which point targets and extended features are displayed on the radar image relative to their true location on the ground. The radar will be designed to generate an image with a constant scale factor in along-track and in cross-track. Furthermore, point targets and other features should be displayed in such a manner that their angular relationship is preserved.

Errors in input signals to the radar can result in degradation of these measures. An incorrect groundspeed of the vehicle carrying the radar will result in an incorrect along-track scale factor. An incorrect altitude input will result in a cross-track scale factor variation across the swath, because the radar collects signals in slant range, but the image needs to be recorded in ground range. Angular image errors—in addition to uneven scale factors—are caused when the image processor displays range lines at some nominal (design) doppler frequency but processes the doppler signals that are centered on another doppler frequency. Moreover, because differences in doppler frequencies represent look angle differences, angular image distortions will result.

1.4 NUMERICAL EXAMPLE

1.4.1 Radar Parameters

To illustrate the performance of a conventional imaging radar and the effects of range and azimuth processing, we postulate a radar with parameters approximately those

Table 1.3
Radar Parameters

Parameter	*Value*	*Dimension*
RF peak power	50	kW
Transmitted pulse width	1	μs
Radar signal bandwidth	15	MHz
PRF	1800	at 900 ft/s
Noise figure	5	dB
Wavelength	0.1	ft
Antenna length	4	ft
Antenna gain	30	dB
Maximum range, high altitude mode	30	nmi
Maximum range, low altitude mode	10	nmi
Grazing angle at 30 nmi, h = 30k ft	9.3	deg
Grazing angle at 10 nmi, h − 6k ft	5.7	deg
Design resolution, range, and azimuth	50	ft

of the AN/APQ-102A radar and proceed to derive the radar performance if that radar were operated in a conventional mode, and then with signal processing. That radar had a maximum range of 30 nmi when operated at high altitude and a maximum range of 10 nmi in the low-altitude modes. In addition, the radar had a fixed-target imaging (FTI) mode and a moving-target imaging (MTI) mode. The radar will be evaluated at the maximum ranges in high and low altitudes, operating in FTI. Clearly, a conventional radar intended for military reconnaissance would use a longer antenna and higher peak power to achieve acceptable performance. The performance of this radar in a conventional mode is derived merely for reference purposes. The radar parameters are listed in Table 1.3.

1.4.2 Performance as a Conventional Radar

A conventional radar that has a radar bandwidth of 15 MHz would not transmit a 1-μs pulse. Therefore, two sets of values are derived: one with a 1-μs pulsewidth and 1-MHz bandwidth, and the other with an 0.066-μs pulse and 15-MHz bandwidth. The results are given in Table 1.4.

We find that such a system yields a high clutter-to-noise ratio even for terrain with a low reflectivity, which is desired in imaging radars, to detect such features as concrete runways in grass areas, for instance. But we also find that the resolution is many times the dimension of such features of interest, and even though the C/N ratio is quite high, the radar image could not display the runways. The signal-to-noise ratio for relatively weak targets, especially in the mode that transmits the higher RF energy, is also quite good, but here we find that the radar could not detect small targets, even against light clutter, as shown in the signal-to-clutter row of the table. The equivalent radar cross section of clutter is the product of the radar backscatter coefficient times the resolution area. Note that the thermal noise power, which is the reference for the scatterometric performance of the radar, is only a fraction of pico watts.

The signal-to-noise ratio to be used for comparison with that derived from the equation after processing is given in the first set of columns, for 1 MHz and 1 μs: 21.7 dB at 10 nmi and 2.6 dB at 30 nmi since that ratio (see equation (1.39)) is proportional to the transmitted pulsewidth, but not to the transmitted bandwidth. For the clutter-to-noise comparison the single-pulse clutter-to-noise values have to be either 11.76 dB (15:1) lower than those in the 1-MHz columns, to account for the higher thermal noise, or 11.76 dB higher than those in the 15-MHz columns, to account for the wider transmit pulsewidth, because that ratio (equation (1.41)) is inversely proportional to the transmitted bandwidth, and proportional to the transmitted pulsewidth. The clutter-to-noise values to be used are 28.7 dB at 10 nmi and 14.3 dB at 30 nmi.

It should be noted that the peak antenna gain has been used for both ranges; this means that the radar is being evaluated in two different operating modes, one

Table 1.4
Conventional Radar Performance

Parameter	*1 MHz* *1 μsec*		*15 MHz* *0.066 μsec*		*Equation Reference*
Range (nmi)	10	30	10	30	
Range resolution (ft)	496	496	33	33	(1.1)
Azimuth resolution (ft)	1520	4560	1520	4560	(1.4)
System noise temperature (K)	917		917		
Noise power (dBm)	−109		−97.2		
Transmitted RF energy per pulse (W sec)	0.05		0.0033		
Per pulse signal-to-noise ratio, σ = 100 ft^2 (dB)	21.7	2.6	9.9	−9.2	(1.7), (1.9)
Per pulse clutter-to-noise ratio, σ_0 = −20 dB (dB)	40.5	26.1	16.9	2.6	(1.8), (1.10)
Equivalent radar cross section of clutter (ft^2)	7539	22610	502	1507	
Signal-to-clutter ratio (dB)	−18.8	−23.5	−7.0	*	

*Signal below thermal noise.

with a maximum range of 10 nmi and the other with 30 nmi. Imaging radars are designed to yield near-constant performance over an entire imaging swath, so if the radar were operated in a mode with a swath extending from 10 to 30 nmi, then the performance at 10 nmi would be close to that at 30 nmi, because the antenna gain in that direction would be lower.

1.4.3 Performance with Coherent Signal Processing

The AN/APQ-102A radar used a chirp network of discrete components (rather than acoustic dispersive delay lines) to generate and compress the transmitted radar signal, which had a linear group delay versus frequency characteristic. The range compression ratio $\gamma_r = \tau_t B_r$ is 15, and after-range compression the ground range resolution is 33 ft at 10 nmi and 30 nmi, which is identical to that of the conventional system, had the radar transmitted the compressed pulsewidth. The radar used optical signal processing, which included doppler filtering, but no resampling. In addition to the number of samples processed by the optical signal processor (filtered, not resam-

Table 1.5
Azimuth Samples

Parameter	*Range* 10 nmi	*Range* 30 nmi	*Equation Reference*
Azimuth beamwidth	1.43°	1.43°	(1.16)
Maximum synthetic-aperture length (same as conv. resolution) (ft)	1520	4560	(1.19)
Processed synthetic-aperture length (ft)	101	303	
Minimum PRF at 900 ft/s	450	450	(1.23)
Selected PRF	1800	1800	
Oversampling factor	4	4	(1.28)
Best azimuth resolution (ft)	2	2	(1.20)
Processed azimuth resolution (ft)	30	30	
Synthetic-aperture time:			(1.25)
for best resolution (s)	1.69	5	
for processed resolution (s)	0.11	0.33	
Maximum illuminated bandwidth (Hz)	225	225	(1.26)
Processed bandwidth (Hz)	30	30	(1.26)
Number of azimuth samples collected from each target			
Actual	3042	9126	(1.24)
Minimum required (n_c/K_a)	760	2281	
Number processed			
Filtered and resampled, minimum	4 (3.34)	10	(1.27)
Filtered, not resampled, maximum	202	606	

pled), Table 1.5 lists the actual and minimum number of samples that need to be collected, and the minimum number of samples to be processed (the azimuth compression ratio). Table 1.6 summarizes the radar performance after signal processing.

Table 1.6
Radar Performance After Signal Processing

Parameter	*10 nmi*	*30 nmi*	*Remarks*
Average trans. power (W)	90		
Range resolution (ft)	33	33	50 or better with errors
Azimuth resolution (ft)	30	30	50 or better with errors
Number of samples processed	202	606	
Signal-to-noise ratio (dB)	44.7	30.4	Equation (1.39) $\sigma = 100$ ft^2
Clutter-to-noise ratio (dB)	34.7	20.4	Equations (1.41), (1.42)
Equivalent cross section of clutter (ft^2)	9.9	9.9	$\sigma_0 = -20$ dB
Signal-to-clutter ratio (dB)	10	10	

This radar has a high clutter-to-thermal noise ratio and will therefore differentiate between no-return areas and light clutter. Furthermore, because the resolution element size is a fraction of the clutter dimensions of interest (runways in grass), it will display such features in the radar image. Note also that after signal processing the signal-to-clutter ratio is positive, which means that the radar will detect small targets in clutter. Fine resolution not only adds detail to the radar image, but also improves the detectability of small targets in clutter.

REFERENCES

[1] *IEEE Standard Dictionary of Electrical and Electronic Terms,* IEEE Std 100, 1988.

[2] Barton, David K. *Modern Radar System Analysis*. Norwood, Mass.: Artech House, 1986.

[3] Klauder J. R., et al. "The Theory and Design of Chirp Radars." *Bell System Technical Journal* **39** (July 1960).

[4] Cook, Charles E., and Bernfeld, Marvin. *Radar Signals*. Academic Press, 1967.

[5] Caputi, W. J. Jr. "Stretch: A Time-Transformation Technique," *Radars: Volume 3, Pulse Compression*. Norwood, Mass.: Artech House, 1975.

[6] Barton, David K., and Ward, Harold R. *Handbook of Radar Measurement*. Englewood Cliffs, NJ: Prentice-Hall, 1969.

[7] Elachi, Charles. *Spaceborne Radar Remote Sensing: Applications and Techniques*. IEEE Press, 1988.

[8] Skolnik, Merrill. *Introduction to Radar Systems*. New York: McGraw-Hill, 1980.

[9] Porcello, Leonard J., et al. "Speckle Reduction in Synthetic Aperture Radars." *J. Opt. Soc. Am.* **66**, no. 11 (Nov. 1976).

[10] Belanger, F. Retired from Goodyear Aerospace Co., Arizona division. Private communication.

Chapter 2

Electronic Warfare Against Imaging Radars

2.1 ELECTRONIC WARFARE DEFINITIONS

The following definitions have been excerpted from Johnson [1].

2.1.1 Electronic Warfare (EW)

Electronic warfare (EW) can be defined as military action that involves the use of electromagnetic energy to determine, exploit, reduce, or prevent hostile use of the electromagnetic spectrum and that retains friendly use of the electromagnetic spectrum. There are three divisions within electronic warfare: electronic warfare support measures (ESM), electronic countermeasures (ECM), and electronic counter-countermeasures (ECCM).

2.1.2 Electronic Warfare Support Measures (ESM)

ESM involves actions taken to search for, intercept, locate, and identify radiated electromagnetic energy for the purpose of immediate threat recognition. Thus, electronic warfare support measures provide a source of information required for immediate action involving electronic countermeasures, electronic counter-countermeasures, avoidance, targeting, and other tactical employment of forces.

An *intercept receiver* (or *search receiver*) is a special calibrated receiver that can be tuned over a wide frequency range to detect and measure radio signals transmitted by the enemy.

2.1.3 Electronic Countermeasures (ECM)

ECM involves the following actions taken to prevent or reduce an enemy's effective use of the electronic spectrum.

- *Jamming* is defined as (1) a form of ECM in which noise or noiselike signals are transmitted at frequencies in the receiver bandpass of the radar to obscure the radar return signal, and (2) the deliberate radiation, reradiation, or reflection of electromagnetic energy with the objective of impairing the use of electronic devices, equipment, or systems by an enemy. (To be effective, a jammer will try to simulate the mode of the radar to be jammed.)
- *Active jamming* is the intentional radiation or reradiation of electromagnetic waves with the objective of impairing the use of a specific portion of the electromagnetic spectrum.
- *Noise jamming* is direct (straight) AM or FM noise on a carrier frequency that has a highly variable bandwidth for the purpose of increasing (saturating) the radar receiver's noise level. Noise jamming is extremely effective against most radars. It may be swept through a frequency spectrum at various rates or spot speeds. Noise jamming is identified by video saturation at all ranges in a sector of the scope; the size of the sector is dependent upon the power and range of the jammer. Random noise may be amplified and radiated directly, or a carrier may be modulated with noise.
- *Barrage jamming* is simultaneous electronic jamming over a broad bank of frequencies.
- *Repeater jamming* involves equipment used to confuse or deceive the enemy by causing his equipment to present false information. This is done by a system that intercepts and reradiates a signal on the frequency of the enemy equipment, the reradiated signal being modified to present erroneous data on azimuth, range, number of targets, and so on.
- *Pulse-modulated jamming* is the use of jamming pulses of various widths and repetition rates.
- *Sidelobe jamming* is done through the sidelobes of the receiving antenna in an attempt to obliterate the desired signal received through the mainlobe of the receiving antenna at fixed points.
- *Passive jamming* is a technique of deception that employs confusion reflectors to return spurious and confusing signals to the transmitting radar set.
- *Confusion reflectors* reflect electromagnetic radiation to create echoes for confusion against radar, guided missiles, and proximity fuses. Confusion reflectors simulate radar returns from targets of interest.
- *Chaff* is an airborne cloud of lightweight reflecting objects typically consisting of strips of aluminum foil or metal-coated fibers that produce clutter echoes in a region of space. Chaff obscures radar returns from targets of interest.

2.1.4 Electronic Counter-Countermeasures (ECCM)

ECCM involves actions taken to ensure friendly effective use of the electromagnetic spectrum despite the enemy's use of EW.

2.1.5 Radar Reconnaissance

Radar reconnaissance is conducted to obtain information on enemy activities and to determine the nature of terrain.

2.2 ELECTRONIC COUNTERMEASURES

Ground imaging radars generate two-dimensional records of the radar reflectivity of the area being imaged. The resultant radar image of extended features (which would constitute clutter in a radar designed to detect specific objects) provides valuable information about the terrain, and the images of point-reflectors provide information on manmade objects. The objective of an imaging radar is to provide such reconnaissance or surveillance data of selected areas, and the objective of jamming against the radar is to deny, confuse, or delay such intelligence of the selected area. Whereas the effectiveness of a jammer against a search or tracking radar is measured in the reduction of the detection or tracking range of the radar, it is the area over which the jammer denies reconnaissance information that is the measure of the imaging radar's susceptibility to jamming.

A jammer may employ different modes to counter an imaging radar. When the radar is known to have frequency agility or frequency diversity capability, barrage jamming is the most effective mode. By radiating jammer noise over a wide bank of frequencies, the jammer is assured that some of the jammer noise will be in the passband of the radar, raising the noise floor and thereby obscuring targets or features of interest. But the available effective radiated power (ERP) of the jammer is now spread over a wide frequency range, which reduces the power spectral density and therefore the jammer noise power within the radar's receiver passband.

Spot jamming is the most effective mode when the victim radar is known to operate on a fixed frequency. As long as all of the jammer noise is within the passband of the radar, the noise level in the radar image is governed only by the ERP of the jammer.

Pulsed jammers are designed to create a large number of false targets in conventional radars but will only raise the noise floor in synthetic-aperture radars, and it is the average ERP of the jammer that determines the noise level in the radar image.

A jammer may employ a deception mode to create false targets in addition to the real targets that are to be protected. To generate deception signals that compress in range and cross range, however, requires that the jammer retransmit an exact replica of the intercepted radar signal and maintains that waveform fidelity for at least one synthetic-aperture time.

2.2.1 Signal, Thermal Noise, and Jammer Noise Spectra

In the time domain, the radar return signal from each pulse is recorded as one line, and that record contains the intrapulse characteristics of the transmitted pulse. The spectrum of that line, typically extending over tens or even hundreds of megahertz, determines the range resolution, and range processing of the line compresses each scattering source into a range-resolvable element. Subsequent radar returns, coinciding with the location of the synthetic-aperture element along the flight path, are recorded in adjacent lines, and a range-resolvable element in that series of lines contains the phase history of any target in that element or cell. The change of phase from pulse to pulse is the doppler shift of the radar return, and that spectrum (typically on the order of kilohertz) extends over a range of ±PRF/2 and determines the cross-range resolution.

Figure 2.1 illustrates the signals in the time domain and the frequency domain. The radar return shown is from a single isolated point target, with the linear FM waveform demodulated against the stable radar reference to video frequencies. The signal frequency sweeps from $-B_r/2$ to $+B_r/2$, is 0 in the center of the radar return, and extends over one transmitted pulsewidth. The spectrum is shown to extend from $-B_r/2$ to $+B_r/2$.

In the first pulse of Figure 2.1(a) the radar return is shown to be in phase with the reference, which results in a maximum at that part of the pulse. The next pulse is shown to be 180° out of phase with respect to the reference, and the radar return coinciding with the center frequency of the radar is a minimum.

The shape of the nth pulse is typical for returns that are 90° out of phase. The values of the radar return at the center of each pulse are plotted below the radar returns (Figure 2.1(b)), and because these values relate linearly to the phase of the return, that figure is also a plot of the phase of the return, as a function of the along-track time. The change of phase as a function of the along-track time is the doppler history of that target; that spectrum covers a range of $\pm B_d/2$. In optically recorded signals, the two time-domain records are orthogonal to each other, and their spectra can also be viewed in two orthogonal dimensions (Figure 2.1(c)).

The two-dimensional spectra should not be confused with the radar ambiguity function of range-doppler search and tracking radars.

Data Format

Because the azimuth processor requires complex samples (magnitude and phase), such samples need to be derived from unambiguous spectra. Both the range and

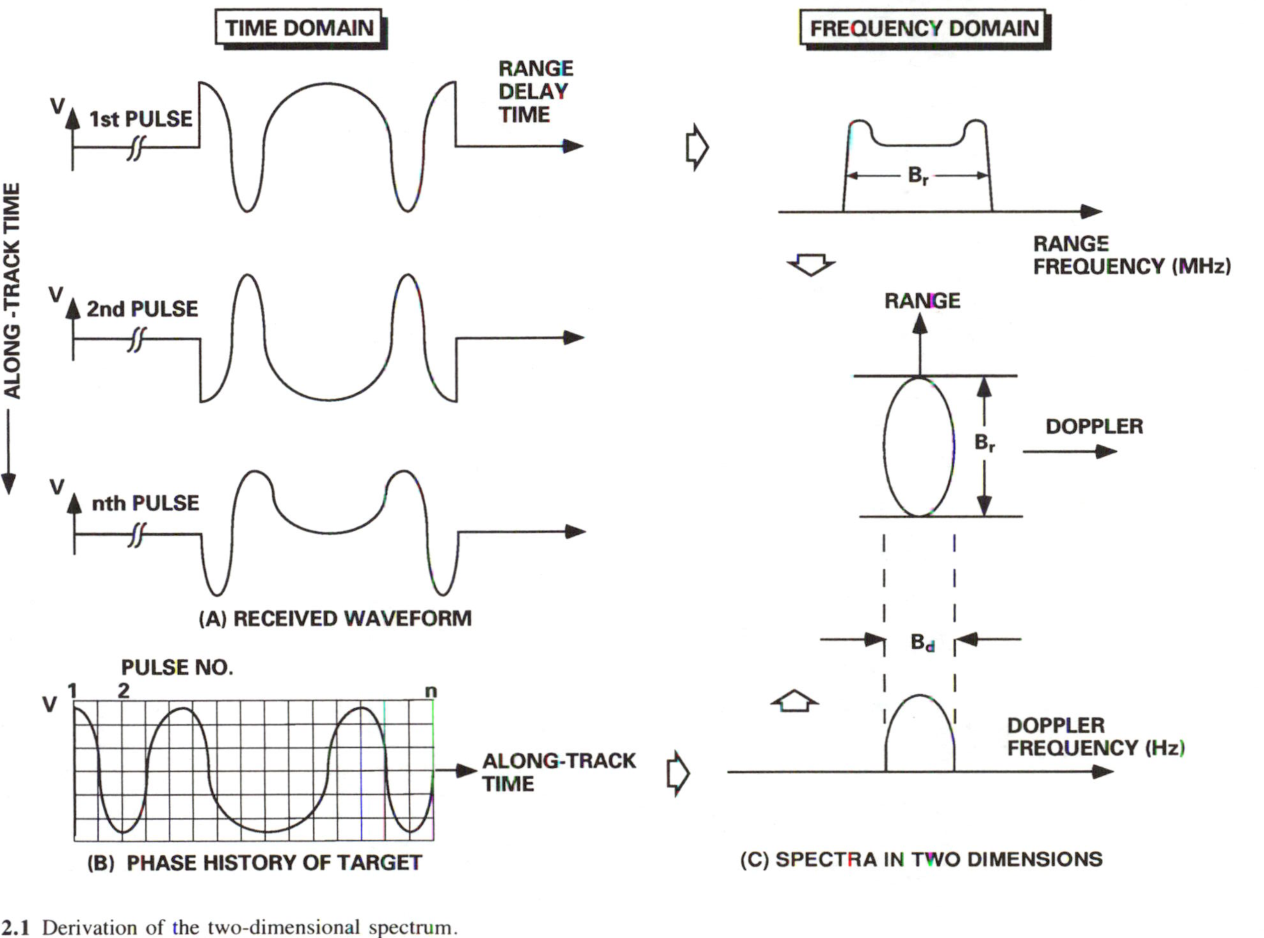

Figure 2.1 Derivation of the two-dimensional spectrum.

doppler spectra of Figure 2.1(c) are ambiguous. To generate unambiguous spectra, the radar data can be stored in an I and Q format, an azimuth-offset format, or a range-offset format.

In-Phase and Quadrature (I and Q)

Figure 2.2(a) illustrates the I and Q format, which is typically used when the data is digitally processed. Thermal noise, as well as spot jammer noise centered on the radar center frequency is also shown.

Thermal noise is filtered by the radar RF components, and the resultant range bandwidth into the processor is equal to that of the radar range bandwidth. In the doppler frequency domain, thermal noise is limited to ±PRF/2. When pulsed at the minimum PRF, as in this case, thermal noise also extends over the same doppler bandwidth as the radar signal (see Figures 1.4 and 1.5).

To obtain complex values, each sample is demodulated against two references, one that is in phase with the transmitted radar signal and another that is in quadrature (90° phase shifted) with respect to the transmitted signal to generate the quadrature values of each sample [2].

Azimuth Offset

When the operating PRF is at least twice the required minimum PRF, the radar data is confined to ±PRF/4 or less, and it is then possible to offset the doppler spectrum by PRF/4 from 0 doppler. This results in an unambiguous doppler spectrum, and complex samples can be derived. Azimuth offset was commonly used in systems with optical signal processing. Figure 2.2(b) shows the azimuth-offset format, with the operating PRF equal to 4 times the minimum PRF ($K_a = 4$). The doppler spectrum has been offset by PRF/4, and it is observed that the thermal noise (and jammer noise) now covers the entire range of possible doppler frequencies. The range bandwidth remains unchanged.

Range Offset

Figure 2.2(c) displays the spectra resulting from range offset. Here the radar transmits the range bandwidth B_r that is required for the processed range resolution, as in the two preceding cases, but the received signal is downconverted to video frequencies with a reference that differs from the reference used to upconvert the radar signal by $B_r/2$. This results in an unambiguous range spectrum, and complex samples can be derived from the total range bandwidth. The signal range bandwidth remains unchanged, as does the thermal noise bandwidth. Whereas the range bandwidth ex-

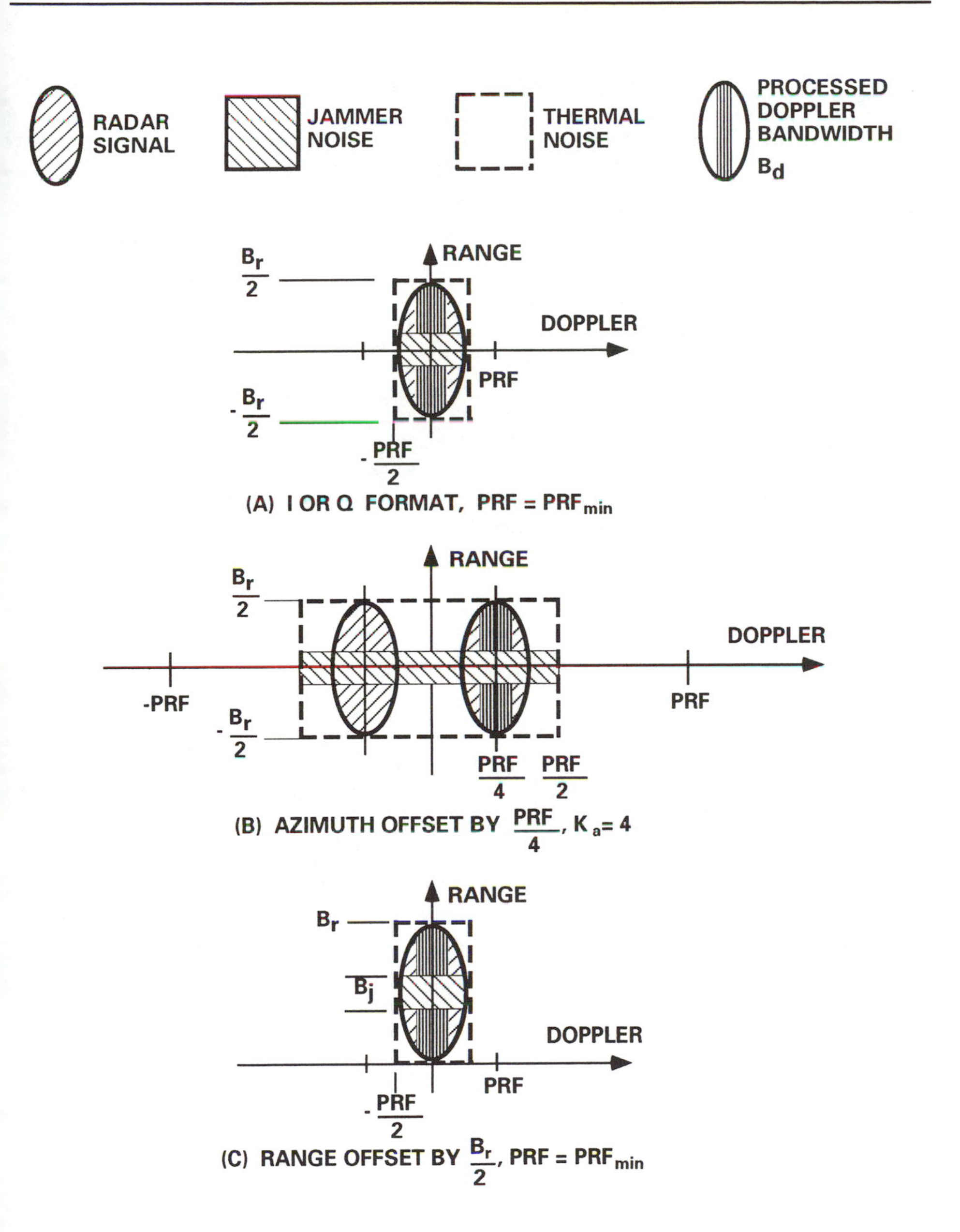

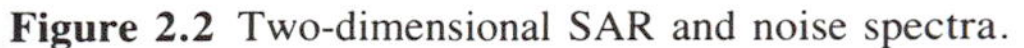

Figure 2.2 Two-dimensional SAR and noise spectra.

tended from + to $- B_r/2$ (for a total of B_r) in the first two cases, it now extends from 0 to B_r, for the same total.

Range offset is not often used, because fine range resolution already requires a wide radar range bandwidth, and range offset doubles the sampling rate requirement of the range bandwidth.

It is interesting to note that of the three methods described for generating complex samples, only azimuth offset provides an improvement in the signal-to-noise ratio, and that improvement is achieved by processing more than the minimum number of azimuth samples.

2.2.2 Jammer Effectiveness

The effectiveness of barrage, spot, random-pulse, and repeater jamming, as well as the image of jammer noise in the radar record (what jammer noise looks like in the radar map) will be discussed in the following paragraphs.

2.2.2.1 Geometry

The geometric relationship between the imaging radar, the swath being imaged, and the jammer location is shown in Figure 2.3. The sensor aircraft flies at altitude h along a flightpath and images a swath in a side-looking mode parallel to the flight path. The azimuth pattern of the radar antenna includes the mainlobe and some sidelobes, and the solid line through the mainlobe identifies the location of the principal elevation plane. Targets illuminated by that part of the antenna pattern will later be processed to form a range line. As the aircraft flies along the flightpath, successive lines are illuminated and subsequently processed, eventually generating a strip map of the terrain.

The jammer is shown to be located in the principal elevation plane when the aircraft is at location "A." The azimuth antenna pattern of the jammer is very wide, so the jammer does not have to track the sensor accurately yet can still radiate the peak ERP in the direction of the sensor.

When the sensor has progressed to position "B," the jammer is in the first sidelobe null of the radar antenna, and the jammer noise intercepted by the radar is a minimum (Figure 2.3(a)). Figure 2.3(b) shows the elevation pattern of the radar antenna, which is usually shaped so as to give constant scatterometric performance over the entire swath.

In the following derivations we will calculate the maximum jammer noise intercepted, which occurs when the peak gain of the radar antenna is pointed at the jammer. The effect of changing radar antenna gains and slant ranges between the radar and the jammer will be addressed in Chapter 3.

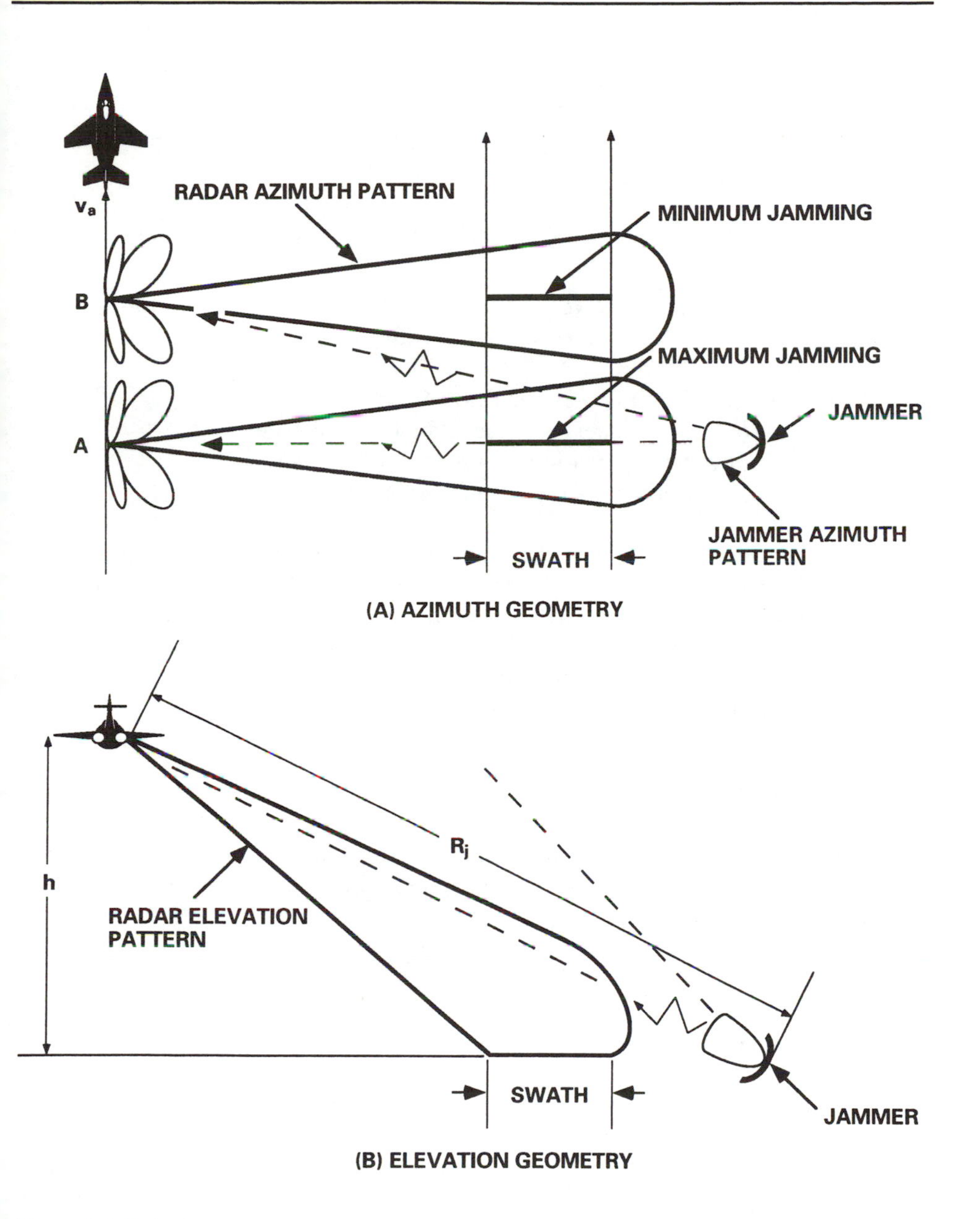

Figure 2.3 Radar-jammer geometry.

2.2.2.2 Intercepted Jammer Noise

The jammer noise intercepted by the radar, regardless of spectral characteristics, is

$$N_{ji} = \frac{P_j G_j}{4\pi R_j^2} A_r = P_J G_J G_t \left(\frac{\lambda}{4\pi R_j}\right)^2 \tag{2.1}$$

which is the jammer noise power density at the radar antenna multiplied by the effective area of the radar antenna in the direction of the jammer. In this case,

P_j = effective transmitted jammer power
G_j = gain of jammer antenna in the direction of the radar
G_t = gain of radar antenna in the direction of the jammer
R_j = slant range between jammer and sensor

We will assume that the gain of the jammer antenna will remain constant during the imaging run of the radar (that is, the jammer is tracking the radar with a wide beam) and that the radar antenna will remain pointed at a constant angle with respect to the velocity vector. The gain of the antenna will change during this time, with a maximum when the jammer is within the elevation plane of the radar antenna (normal to the flight path for the side-looking case).

2.2.2.3 Jammer Noise

Random jammer noise within the processing aperture of $B_r \times B_d$ is noncoherent in range and azimuth and is acted upon by the signal processor in the same manner as thermal noise:

$$N_{jra} = N_{ji} \times K_f \times n \tag{2.2}$$

for linear FM signals processed by dispersive delay lines, and

$$N_{jra} = N_{ji} \times K_f \times \tau_t B_r \times n \tag{2.3}$$

for processing by correlation or FFT (n is the number of coherently processed azimuth samples, as defined in Section 1.2.2). The spectral mismatch factor K_f, which is always less than or equal to 1, limits the intercepted jammer noise bandwidth to that part within the radar receiver bandpass ($K_f = B_r/B_j \leq 1$).

Noise from all sources competes with the signal and clutter on the radar image. Noise can be of a thermal nature, as discussed previously; there may also be noise contributed by ambiguities, as well as quantization and saturation noise, all of which contribute to the noise floor of the radar image. In an EW clear environment, a well-designed radar will yield an adequate signal-to-noise or clutter-to-noise floor ratio. Jammer noise adds to that noise floor, and all sources should be considered when evaluating the effectiveness of jammers. However, the jammer noise will have to exceed the noise floor significantly to be effective in obscuring radar image features, so we can disregard the noise sources contributing to the noise floor and concentrate on the signal-to-jam and clutter-to-jam ratios to evaluate the susceptibility of a synthetic-aperture radar to jamming, as well as the effectiveness of ECCM (see Chapter 3).

To express the signal-to-jam and clutter-to-jam ratios in terms that address the signal processing effects, we use previously derived equations for signal-to-noise and clutter-to-noise and multiply those equations by the thermal noise-to-jammer noise ratio.

$$(S/N_j)_{ra} = (S/N)_{ra}\left(\frac{N_{ra}}{N_{jra}}\right) = \left(\frac{S}{N}\right)_{ra}\left(\frac{k\,T_s\,B_r}{P_j\,G_j\,G_t\,K_f}\right)\left(\frac{4\pi R_j}{\lambda}\right)^2 \tag{2.4}$$

Using equation (1.39) results in

$$(S/N_j)_{ra} = \left(\frac{P_t\,\tau_t\,\text{PRF}\,G_t}{4\pi P_j\,G_j\,K_f}\right)\left(\frac{R_j^2}{R^3}\right)\left(\frac{\lambda\sigma}{2\,v_a}\frac{B_r}{w_a}\right) \tag{2.5}$$

The signal-to-jam ratio improves with the average transmitted radar RF power. The preferred method to achieve the required average power is to use a long pulse or a high PRF, or both. Whereas the high peak power has the same effect, it also raises the visibility of the radar to an intercept receiver and therefore the probability of intercept.

The signal-to-jam ratio in a synthetic-aperture radar also improves with the range compression ratio (increasing the range bandwidth B_r but keeping the transmitted pulsewidth constant) and the azimuth compression ratio, which is reflected in the processed azimuth resolution, w_a.

The clutter-to-jam ratio is derived in the same manner, using equation (1.42):

$$(C/N_j)_{ra} = \left(\frac{P_t\,\tau_t\,\text{PRF}\,G_t}{4\pi P_j\,G_j\,K_f}\right)\left(\frac{R_j^2}{R^3}\right)\left(\frac{\lambda\sigma_0 c}{2\,v_a}\frac{1}{2\cos\theta_g}\right) \tag{2.6}$$

The clutter-to-jam ratio also improves with the average transmitted radar power but is independent of resolution and compression ratios.

2.2.2.4 Deramp-on-Receive

The deramp-on-receive method of signal processing was briefly addressed in Section 1.2.2. Because such processing results in a reduction of the range-related bandwidth from B_r to one that is given by the FM slope of the transmitted pulsewidth times the swath time, and that bandwidth is less than the range bandwidth (which is one objective of deramp processing), it is of interest to assess the effect of deramp processing on jammer noise. Figure 2.4(a) shows the noise bandwidth $B_j < B_r$, such that all the jammer noise is within the passband of the radar. After deramping, the signal bandwidth can be less than the jammer bandwidth (depending on the swath being processed). Figure 2.4(b) shows that all of the jammer power is still present. Furthermore, whereas the jammer noise occupied a fraction of the radar range band-

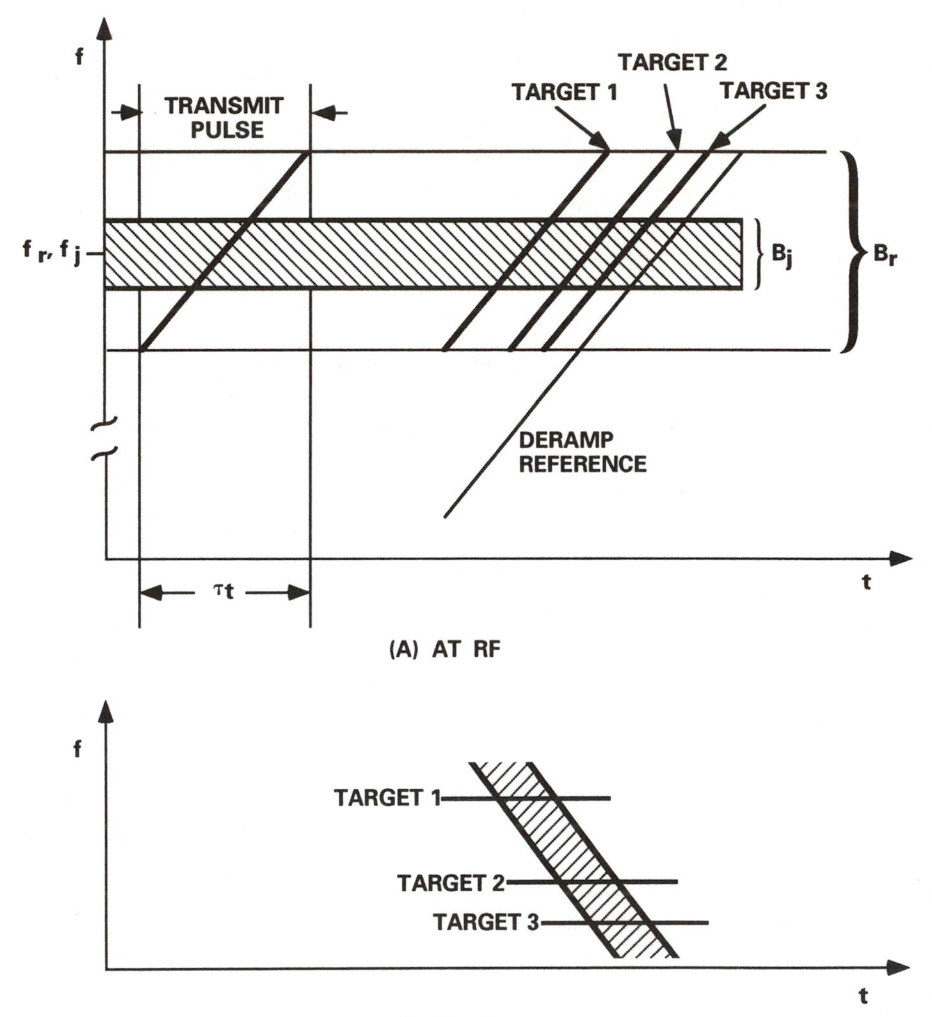

Figure 2.4 Jammer noise in deramp-on-receive.

width prior to processing the noise now occupies the same fraction of the transmitted pulse width.

Note that the deramped signal bandwidth is not to be used to calculate the spectral mismatch factor.

2.2.2.5 Effective Jammer Noise Versus Jammer Mode

Not all of the jammer noise intercepted by the radar contributes to the noise floor of the radar image. Aside from a possible polarization mismatch, which is independent of the jammer mode, the intercepted jammer noise can be reduced by spectral mismatch and by the duty cycle of a pulsed jammer.

Barrage Jamming

A barrage jammer transmits in a continuous mode, and the effective jammer noise power to be used in equations (2.5) and (2.6) is the peak transmitted power. The bandwidth B_j is much wider than the radar bandwidth. The radar RF components reject the jammer noise outside the receiver bandpass, and only the in-band jammer noise can eventually contribute to the noise floor. For narrowband radars, the spectral mismatch factor may be on the order of −20 dB, but for fine resolution, wideband imaging radars is more likely to be on the order of −6 dB.

The jammer noise extends over the entire swath and is of uniform intensity. The radar image of barrage jammer noise will exhibit speckle (see Section 1.2.3). Barrage jammer noise covers the entire radar range bandwidth and the entire range of possible doppler frequencies. The processor limits the doppler bandwidth to that required for the processed cross-range resolution and performs the Fourier transform of the range and azimuth spectra of the jammer noise. The speckle of the processed jammer noise will be on the order of the range and cross-range resolution element dimension. In addition, because a large number of noise samples are added noncoherently, the multiple looks of jammer noise tend to smooth out the intensity variation from pixel to pixel, just as in the case of thermal noise.

Spot Jamming

A spot jammer also radiates noise in a continuous mode, such that the effective transmitted jammer power is the peak power. However, the jammer noise spectrum will be fully contained within the radar receiver bandpass ($B_j < B_r$), and the spectral mismatch factor is 1. (Even though the equation for K_f given in Section 2.2.2.3 can result in values larger than 1 for spot jammers, that factor can assume values only from 0 to 1, which means that there can be no more than all jammer noise within the radar bandpass.)

Spot jammer noise also covers the entire swath and is of uniform intensity across the swath. However, the image of this jammer noise will differ from the barrage noise, because the Fourier transform of the narrower band jammer noise will result in speckle size in the range dimension that is larger than that of thermal noise or clutter. The processed cross-range dimension is again equal to that of clutter or thermal noise. Spot jammer noise will appear to be stretched in range.

Random Pulse Jamming

The effective jammer power for barrage and spot jammers was equal to the peak transmitted power. For random pulse jamming, however, the effective jammer power is the average transmitted power. Because the jammer pulses are transmitted at random intervals, such noise pulses can appear in any part of the range swath. When observed over a sufficient number of samples (radar returns in the time domain), the noise pulses will occupy all parts of the range swath in one sample or another. The azimuth processor forms the sum of the noise power from all samples within one synthetic-aperture length. That sum will be equal to the total noise power in the aperture, which is proportional to the average jammer noise power (see also the discussion of processing of weighted noise in Section 1.2.3).

The range bandwidth of the pulsed jammer will be within the radar range bandwidth; the spectral mismatch factor is again 1. The jammer noise will extend over the entire range swath, and the speckle dimension will again appear stretched in range, just as in the spot jammer case. However, the random pulse jammer speckle will exhibit more pronounced brightness variations than that from spot or barrage jamming, because fewer noise samples are added noncoherently, thereby reducing the smoothing effect of multiple looks (see Section 1.2.3). (Random pulse jamming has the effect of generating a large number of false targets in conventional radars but generates only noise in synthetic-aperture radars.)

Repeater Jamming

Depending on the fidelity with which the repeater jammer retransmits the intercepted radar signal, such jammer signals can generate noise in a number of range cells, or can generate false targets. If the coding of the radar pulse and the doppler characteristics are maintained, then such signals will compress in range and cross range, and the intensity of the false targets is enhanced by the signal processing gains, just like radar returns from fixed point targets.

When the radar pulse characteristics are not maintained in the transponding process, the signal processor will act on the repeater signals in the same manner as on noise. The effective transmitted power is the peak power of the jammer. Furthermore, because the transponded signal will resemble the intercepted radar signal, the spectral mismatch factor is 1 ($B_j = B_r$). In the range direction, the jammer noise will be confined to a number of range cells, and that width is governed by the width of the retransmitted jammer pulse. The location of the jammed range cells is related to the location of the jammer and the time delay introduced by the repeater. In this case, the radar image of the jammer noise will resemble that of barrage noise.

When the characteristics of the radar signal are preserved in the transponded signal, it is no longer proper to refer to the processed jammer pulses as noise. Assuming that only the intrapulse characteristics (pulsewidth, bandwidth, and FM linearity) are preserved, then the pulses will compress in range. In the absence of doppler fidelity, the resultant radar image of the compressed pulses is a line with the width of a compressed range resolvable element instead of the narrow noise strip. The intensity of that line will be enhanced by the range compression ratio γ_r, compared to the intensity of the noise strip. At that signal level the sidelobes of the compressed range response may become visible and be displayed as faint lines parallel to the main response.

If the repeater also preserves the phase of the intercepted radar signal on a pulse-to-pulse basis, then the transponded pulses will also compress in cross range. The result is an image of a point target, and the intensity is further enhanced by the azimuth compression ratio γ_a. Such targets can be expected to be very strong and would be associated with range and cross-range sidelobes of the processed false targets. In addition to the image of the false target that was received through the main lobe of the azimuth antenna pattern, there may be ambiguous false targets generated at along-track locations in which the doppler frequency of the transponded signal is equal to the PRF of the radar. Such signals would be received through the sidelobes of the radar antenna.

Using the numerical example of Section 1.4, we find that the width of the jammed part of the swath is approximately 500 ft. The width of the compressed range line would be 33 ft, and the intensity of that line would increase by 11.7 dB. The intensity of the false point target would increase an additional 23 dB or so.

It becomes evident that it does not require much jammer power to generate false targets, but the complexity of a high fidelity repeater jammer is significant. An alternate (and much cheaper) method of creating false targets in a radar image is to position corner reflectors (passive jamming) at desired locations.

2.2.2.6 Along-Track Extent of Jamming

The obscuration of targets and features of interest was shown to cover the entire swath for barrage, spot, and random-pulse jamming but only a narrow strip within the swath for repeater jamming. In along track, the obscuration will extend over a distance that is governed by the one way azimuth pattern of the radar antenna and the jammer ERP. If the jammer noise obscures only targets and features of interest when the jammer is in the mainlobe of the radar antenna, that along-track distance is

$$D_j = R_j \lambda / L \tag{2.7}$$

where R_j is the slant range distance between the jammer and the sensor. However, conditions can also be such that the jammer noise can deny reconnaissance data when the jammer noise is received through the sidelobes of the antenna pattern. The radar image will exhibit minimal jamming when the jammer noise is received through the sidelobe nulls and effective jamming when received through the sidelobe peaks.

2.3 JAMMING OF DATA-LINKED SIGNALS

Imaging radars may employ data links to provide the desired information to ground-based users in near-real time. Theoretically such a link is susceptible to jamming, because the signals make use of the RF spectrum. However, examining the signal flow and the geometry reveals that data links are not really vulnerable to jamming.

The radar data originates at the sensor and is transmitted to the ground station. A jammer would have to intercept that signal and transmit noise in the direction of the ground station, the location of which is not necessarily known. Even if the ground station radiates up link signals that could be intercepted and thereby reveal the location, there is still the problem of the sensor—data link—jammer geometry. The radar will image a swath to one side of the flight path and transmit the data to the other (friendly) side. Assuming that the jammer managed to intercept the down-linked signal and fix the location of the ground station, the jammer would have to transmit from a location above the radar horizon of the ground station. Furthermore, because ground stations track the sensor accurately, they are associated with very narrow beams, and a jammer would have to be almost in line with the sensor and the ground station to be effective. One concludes that data links used by imaging radars are virtually immune to jamming.

2.4 ELECTRONIC COUNTER-COUNTERMEASURES

2.4.1 General

Jamming effectiveness is measured as the area in which radar reconnaissance or surveillance data is denied or impaired. The effectiveness of ECCM is measured as the reduction of that area. To initiate jamming, a jammer must have knowledge of the radar's presence, which can be obtained by intercepting the radar's transmitted signal. The intercepted signal must be classified to assess the threat, by measuring the radar parameters. In case of multiple threats, priority has to be established and

ECM resources allocated, so that the most effective jammer mode against the threat can be selected.

Electronic counter-countermeasures are designed to reduce the effectiveness of jamming. The initiation of jamming can be delayed (and, therefore, the along-track extent of the jammed area reduced) by means that reduce the radar signal power density at the intercept receiver location. The initiation of jamming may also be delayed by changing the radar parameters used for classification or by the transmission of deception signals by the radar. This creates uncertainty about the type of electronic threat (classification) and lengthens the jammer's reaction time. In addition, the level of jamming can be reduced by means that act to suppress the jammer noise without degrading the desired radar return signal significantly.

In the following paragraphs we will address the subjects of radar signal intercept, radar signal classification, and jammer noise reduction.

2.4.2 Radar Signal Intercept

The radar RF power intercepted by a receiving station is

$$P_{ri} = \frac{P_t G_t}{4\pi R_i^2} A_i = P_t G_t G_i \left(\frac{\lambda}{4\pi R_i} \right)^2 \tag{2.8}$$

where

A_i = effective area of intercept receiver antenna
G_i = gain of intercept antenna
R_i = range to the intercept receiver

The gain of the intercept antenna will be assumed to remain constant, and the gain of the radar antenna will vary during an imaging run, with a maximum when the interceptor is in the principal elevation plane of the radar antenna pattern. When that intercepted radar power—normalized to the receiver's thermal noise and enhanced by any processing gain that the intercept receiver may have—exceeds some threshold, then the presence of the radar will be known.

$$\frac{P_{ri}}{N_i} \gamma_i = \left(\frac{P_t G_t G_i}{k T_s B_i} \right) \left(\frac{\lambda}{4\pi R_i} \right)^2 \gamma_i > \left(\frac{S}{N} \right)_i \tag{2.9}$$

where

$N_i = KT_sB_i$ = thermal noise level of intercept receiver
B_i = instantaneous receiver bandwidth
γ_i = intercept receiver processing gain

The signal-to-noise ratio, $(S/N)_i$, to be used in equation (2.9) depends on the probability of intercept that is required and the false alarm rate than can be tolerated.

When evaluating the probability of intercept, it is of interest to express the radar ERP in terms of the other parameters of equations (1.39) and (1.42). Substituting in equation (2.9) results in equations (2.10) and (2.11).

$$\frac{P_{ri}}{N_i}\gamma_i = \left(\frac{S/N}{\sigma}\right)\left(\frac{1}{\tau_t B_r}\frac{2\,v_a\,w_a}{\text{PRF}}\gamma_i\frac{B_r}{B_i}\right)\left(\frac{G_i}{G_t}\frac{R^3}{R_i^2}\frac{4\pi}{\lambda}\right) \tag{2.10}$$

$$\frac{P_{ri}}{N_i}\gamma_i = \left(\frac{C/N}{\sigma_0\,w_r\,w_a}\right)\left(\frac{1}{\tau_t B_r}\frac{2\,v_a\,w_a}{\text{PRF}}\gamma_i\frac{B_r}{B_i}\right)\left(\frac{G_i}{G_t}\frac{R^3}{R_i^2}\frac{4\pi}{\lambda}\right) \tag{2.11}$$

Equations (2.10) and (2.11) show that the probability of detection of the radar by an intercept receiver is governed by the following:

- *Reconnaissance objective* (first term of the equations): Radars that produce high S/N ratios for targets with low cross sections, and high C/N ratios for low reflectivity areas imaged with a high resolution (that is, radars that produce high-quality images) are more vulnerable to intercept.
- *Low probability of intercept (LPI)* [3] (second term of the equations): Radars that employ high processing gains (discussed in Section 1.2) that are not compensated by intercept receiver processing gains are less vulnerable to intercept. The probability of intercept can also be reduced if the intercept receiver can be forced to a wide instantaneous bandwidth, B_i.
- *Radar and jammer range and intercept antenna gain (pattern)* (third term of the equations): The effect of range and antenna gains cannot be easily visualized because of the multitude of possible combinations. The interaction of range and antenna patterns will be addressed in detail in Chapter 3.

Processing gains for the multitude of receiver types and possible waveforms are not derived; the readers are merely alerted to the fact that intercept receivers may employ signal processing to enhance the intercepted radar signal.

2.4.2.1 Radar ERP

The radar power intercepted by all types of intercept receivers is proportional to the effective radiated power (ERP) of the radar, which is the product of the peak transmitted radar power and the gain of the transmitting antenna. Reducing the ERP of the radar is the most effective method against intercept, but to achieve constant performance for configurations that use lower ERP, the system needs to employ higher processing gains. There are limits to this approach, imposed by timeline considerations (maximum pulsewidth, maximum PRF) as well as the complexity of signal processors. In equation (1.39), the range processing gain is reflected in the transmitted pulse width, τ_t, with the radar bandwidth remaining constant for constant range resolution. The azimuth processing gain is reflected in the term $\text{PRF}/2w_a v_a$, which is proportional to the azimuth compression ratio, γ_a.

Low probability of intercept (LPI) is achieved by spreading the radar return signal that is required for a given radar S/N ratio over a wide pulsewidth and over a large number of pulses in such a manner that the intercept receiver processing gains are poorly matched to the radar signals. In examining the second terms in equations (2.10) and (2.11), it becomes evident that high processing gains alone do not necessarily result in a low probability of intercept. The LPI feature of a waveform can be stated only with respect to a particular intercept receiver type.

Clearly, if an intercept receiver could match all processing gains of the radar, then the probability of detecting the presence of the radar signal would be essentially the same for all configurations that have the same radar performance (product of radar ERP and processing gains) [3].

The radar will be dimensioned to yield the desired performance at the maximum operating range and, within the design constraints, will result in some radar ERP. In a multimode radar in which the modes differ in the imaging range and resolution, this may result in excess RF power at the shorter range modes, and the peak power can then be reduced for those modes without any sacrifice in imaging performance. This is one form of RF power management.

When operating in a known EW environment, it may be desirable to trade poorer image quality for a lower probability of intercept. In that case, the radar peak power is lowered to achieve a desired signal-to-noise ratio only for strong targets. This is sometimes referred to as highlight imaging, or operating in a quiet mode.

2.4.2.2 Antenna Pattern

The desired radar performance is also obtained by the contribution of the product of the transmitting and receiving antennas gains. Using a lower-gain transmitting antenna (compensated by a higher-gain receiving antenna) will reduce the power density at the intercept receiver site, but this will result in an increased vulnerability to jamming. Low sidelobe antennas reduce the probability of intercept and the vulnerability to jamming in the sidelobe region; this is the preferred antenna pattern.

2.4.2.3 Center Frequency

Frequency agility [1, 3, 4, 5, 6] (changing the center frequency of the radar on a pulse-to-pulse basis) is not an option for a synthetic-aperture radar. The system has to dwell on a constant center frequency for at least one synthetic-aperture time t_s. If the operating frequency is changed during an imaging run, a discontinuity as well as a change in the slope, or focal length of the phase history of the illuminated targets, results (measuring phase is equivalent to measuring distance differences from pulse to pulse to within a fraction of a wavelength), and the processed radar image will exhibit degradations of the cross-range resolution. The resolution of targets illuminated when the frequency was changed will be most severely degraded (by a factor of two), because the phase history will match the reference function—which also may have to be changed—for one-half of the synthetic-aperture length at most. The degradation will extend over an along-track distance of $\pm L_s/2$ from that point but will diminish to zero at both ends of the interval.

Changing the center frequency of the radar in a continuous mode of operation frequently is possible but not practical. However, if the radar is operated in a burst mode [7] (also called a *batch mode*), the center frequency can be changed from burst to burst without any degradation in image quality. (The radar is operated for one synthetic-aperture time of the processed cross-range resolution, which has to be coarser than the best obtainable resolution in a continuous mode, and the excess doppler bandwidth is processed to form additional adjacent cross-range lines, thereby generating an entire scene.) The burst mode results in intermittent radar operation with short radar "on" times and is effective in delaying the start of jamming when combined with frequency diversity.

2.4.3 Radar Signal Classification

After the radar signal has been detected, that signal needs to be classified to determine the threat that that signal represents. The signal parameters used for classification are the center frequency and bandwidth (spectral), the pulse modulation (intrapulse features), and PRF (interpulse features). The classification of the threat could be delayed—or even denied—if the parameters used for classification are changed during the classification process. The following is an assessment of the potential to change such parameters.

2.4.3.1 Center Frequency

Frequency diversity is possible in the burst mode of operation, as discussed in Section 2.4.2.3.

2.4.3.2 Radar Bandwidth, Pulsewidth, and Pulse Modulation

Pulse compression is accomplished on a pulse-to-pulse basis and, except for the ease of mechanization, is independent of doppler processing. Therefore, these parameters could be changed on a pulse-to-pulse basis with the intent of delaying or denying classification. However, any such change will certainly complicate the range processor. In addition, changing the radar bandwidth during a synthetic build-up time will tend to degrade the image quality in range, such as resolution and sidelobes. Changing the pulsewidth from pulse to pulse, without a corresponding change in bandwidth, will change the time-bandwidth product and introduce amplitude errors over the synthetic aperture. This will tend to degrade the image quality in cross range. Changing the pulse modulation on a pulse-to-pulse basis, while holding the bandwidth and the pulse compression ratio constant, will not degrade the image quality.

2.4.3.3 Interpulse Period

The interpulse period fixes the location of the phase center of the physical antenna along the flight path, and each transmission of a pulse becomes an element of the synthetic array. A constant PRF at a constant vehicle velocity results in a uniform spacing of such array elements. In addition, just as it is not necessary to have a uniform spacing of radiating elements in a physical antenna, it is not necessary in a synthetic antenna, as long as the spacing is no larger than one-half of the length of the physical antenna and the true location of each element is known. A simpler implementation of changing the interpulse period could be to drop pulses occasionally; this would result in a thinned array, with some degradation in the cross-track image quality.

2.4.3.4 Deception Signals

Deception signals—which differ from the radar signal primarily in the center frequency but can also differ in bandwidth, PRF, and the type of modulation—can be transmitted to counter classification. This requires either a separate source of RF power or excess RF power of the radar. The burst mode of operation lends itself to the transmission of deception signals, because the radar is operated for only one synthetic-aperture length and is off for the remainder of the flightpath equal to the processed image scene. Deception signals can be transmitted during this "off" time. To provide for a higher visibility of the deception signal, when the radar is viewed through its sidelobes it is advantageous to transmit such signals through a broad auxiliary beam [5].

2.4.4 Jammer Noise Reduction

The effectiveness of the jammer can be reduced even if the jammer operates in the radar operating frequency band. To counter a jammer effectively, it is necessary to exploit characteristics of the jammer signal that differ from those of the desired radar return signals. A differentiation can be made on the basis of spatial characteristics, temporal characteristics, spectral characteristics, and modulation characteristics of the jammer signal. In addition, operational procedures can be taken to reduce the effectiveness of a jammer.

2.4.4.1 Spatial Characteristics

The jammer signal strength received by the radar system is proportional to the gain of the radar antenna in the direction of the jammer and generally is of such magnitude that even when received through the sidelobes of the radar antenna, degradation of the radar image results. The countermeasure to jammer signals received through the sidelobes is to reduce the gain in the direction of the jammer by employing a low-sidelobe antenna, which also reduces the probability of intercept through the side-lobes. The level of the jammer noise can also be reduced by pointing a null in the direction of the jammer. This can be accomplished by tracking the jammer with an existing null of the antenna pattern (azimuth or elevation) while keeping the area to be imaged illuminated by the mainbeam of the pattern, or by generating a null in the direction of the jammer. Sidelobe cancellers and null steering antennas are effective in countering jammer noise received through the sidelobes. Sidelobe blanking is effective in radars that scan rapidly over the jammer site (scanning antennas) but not practical in synthetic-aperture radar that sweep gradually over the jammer locations. These techniques are defined in the *IEEE Standard Dictionary*.

The improvement in signal-to-jam ratio is directly proportional to the reduction of the antenna gain in the direction of the jammer. Although such techniques are primarily applicable to the azimuth sidelobe region of the SAR antenna, jammer tracking in the main lobe with a monopulse null and imaging through the difference pattern are possible. This extends the effectiveness of null steering ECCM devices into the main lobe region.

2.4.4.2 Temporal Characteristics

The time of arrival of radar return signals with respect to the transmitted signal is governed by the minimum and maximum radar ranges that are illuminated by the antenna elevation pattern. Signals received at times other than those expected are not radar returns, and this can be exploited to detect the presence of jamming as well as to measure signal strength and spectral characteristics. The incorporation of a jamming detector and jammer noise analyzer are a necessary addition to any system that incorporates ECCM features.

2.4.4.3 Spectral Characteristics

When the jammer signal occupies only a fraction of the sensor's signal frequency band, selective attenuation of that part of the sensor's frequency spectrum that contains the jammer will improve the radar's signal-to-jam ratio. With notch filtering (defined in the *IEEE Standard Dictionary*), the resultant radar image will exhibit a significant improvement in the signal-to-jam ratio but will also be degraded in resolution, sidelobes, and signal-to-noise ratio. Such notch filtering is most effective at a point in the radar receive signal path before saturation. (For strong jammers this is ahead of the low-noise amplifier.) However, notch filtering at IF can be expected to be effective and has even been found to yield some signal-to-jam improvement at video frequencies in the signal processing path. Notch filters are effective against narrowband jammers (spot jammers) when used in wideband radars and are equally effective against all types of jammer noise modulation.

The notch filter attenuates the jammer noise, and if the band-reject width of the filter equals or exceeds the jammer bandwidth and the attenuation of the filter exceeds the jammer noise to thermal noise ratio, then the jammer noise will be completely eliminated. [Residual jammer noise can remain if the filter does not track the jammer center frequency properly or if the conditions stated above (reject width and depth) are not met.] Even if the jammer noise is completely eliminated, however, the radar image will show degradations in the signal-to-noise and clutter-to-noise ratios, as well as in range resolution and range sidelobe performance. The notch filter attenuates not only the jammer noise, but also a part of the radar signal spectrum. Thermal noise will also be attenuated if the filter is placed after the low-noise amplifier (LNA). In this case the signal and noise level out of the filter will be reduced by the same factor, namely $(1 - B_n/B_r)$, where B_n is the effective reject width of the filter. However, if the filter is positioned in front of the LNA (to prevent saturation), then only the signal level will be reduced, not the thermal noise.

The notch filter will also reduce the pulse compression ratio because it deletes a part of the radar spectrum and, in the case of linear FM, also a part of the pulsewidth. If the band reject is placed at an end of the radar bandwidth, then the remaining bandwidth is $B_r - B_n$, and the remaining pulse width is $\tau_t(1 - B_n/B_r)$, for a resultant pulse compression ratio of $\tau_t\, B_r(1 - B_n/B_r)^2$. It is observed that the signal-to-noise ratio after notch filtering is severely impaired if the reject band of the filter is a significant percentage of the radar signal bandwidth.

Assuming that the AN/APQ-102A radar (see Section 1.4) were to counter a spot jammer with a bandwidth of 5 MHz using a notch filter at RF with a matching reject band, then the resultant signal-to-noise ratio would be proportional to a pulse compression ratio of only 6.6:1 rather than the initial 15:1, and the range resolvable element size would degrade to 50 ft from 33 ft. The best obtainable reject bandwidth of a tracking notch filter of RF is more likely to be greater than 5 MHz. At an optimistic notch bandwidth of 10 MHz, the signal-to-noise ratio would degrade by close to 10 dB and the range resolvable element size would degrade to 100 ft. Clearly the AN/APQ-102A radar was not wideband enough for a notch filter ECCM.

2.4.4.4 Modulation Characteristics

Radar return signals from terrain are primarily amplitude-modulated, and jammer signals will be primarily frequency-modulated to maximize the effective radiated

Table 2.1
Summary of ECCM Options

Parameter	Method	To Counter: Detection	Classification	Noise	Comment
Center frequency	Frequency diversity	X	X		In burst mode
	Deception		X		Burst mode or added RF source
RF peak power	Select for const S/N	X			Programmed
	Reduce for highlight imaging	X			Selectable
	Use maximum power (burnthrough)			X	After detection in low-power mode
Antenna pattern	Low sidelobes	X		X	
	Sidelobe canceller			X	Against single narrowband jammer
	Null steering			X	Limited to number of nulls available
Intrapulse	Change bandwidth		X		Degraded range and cross-range resolution
	Change pulsewidth		X		Degraded range and cross-range resolution
	Change modulation		X		No degradation
Interpulse	Change PRF		X		Increased processor complexity
	Drop pulses		X		Degraded cross-range resolution, sidelobes
Jammer spectrum	Notch filter			X	Against narrowband AM, FM, pulsed jammers
Jammer modulation	Noise canceller			X	Against narrowband and wideband FM, single source jammers
	Limit dynamic range of radar data			X	Against pulsed jammers

power of the jammer. This characteristic is being exploited in noise cancellation schemes in which the jammer signal is subtracted from the composite (radar and jammer) signal. Noise cancellers are effective against frequency-modulated jammers and jamming signals that emanate from a single source. Furthermore, noise cancellers require a jammer amplitude that is large compared to the average radar return signal before they can be engaged.

Pulsed jammers inject noise into the radar image, which is proportional to the average effective radiated power of the jammer, because a large number of radar returns are processed in an image processor to generate a pixel. The effectiveness of a pulsed jammer can be reduced by time-gating the jammer out. Another method of countering a pulsed jammer is to reduce the dynamic range of the received signal by limiting the amplitude of the composite signal.

2.4.4.5 Other Procedures

After the radar has been detected and jamming initiated, while the radar operated with reduced RF peak power (quiet mode, short range modes), the radar can revert to transmitting the maximum available peak power to improve the signal-to-jam ratio. This is called *burnthrough* [6].

REFERENCES

[1] Johnston, Stephen L. *Master Electronic Warfare Glossary*. International Countermeasures Handbook, 1976–1977.

[2] Hovanessian, S. A. *Introduction to Synthetic Array Radars*. Norwood, Mass.: Artech House, 1980.

[3] Schleher, D. Curtis. *Introduction to Electronic Warfare*. Norwood, Mass.: Artech House, 1986.

[4] IEEE Standard Dictionary of Electrical and Electronic Terms, IEEE Std 100, 1988.

[5] Barton, David K. *Modern Radar System Analysis*. Norwood, MA: Artech House, 1986.

[6] Skolnik, Merrill. *Introduction to Radar Systems*. New York: McGraw-Hill, 1980.

[7] Elachi, Charles. Spaceborne Radar Remote Sensing: Applications and Techniques. IEEE Press, 1988.

Chapter 3
Electronic Warfare Effectiveness Measure

3.1 CONTOUR PLOT CONCEPT

3.1.1 General

The vulnerability measure of a synthetic-aperture imaging radar was stated as the imaged area over which a jammer denies certain reconnaissance information. Because the radar will be designed to yield near-constant performance over an entire imaged swath (range dimension), the along-track dimension of the jammed imagery is a valid measure (see Section 2.2.2.6). Similarly, the along-track distance over which the radar remains detected is the measure of the vulnerability of the radar to intercept.

Plots of lines of constant power density on the ground provide such a measure. The plots are unique to mission parameters of the radar, such as altitude, mode, and squint angle—which is reflected in R_o, the slant range at which the boresight axis of the antenna intercepts the ground—and to antenna patterns in azimuth and elevation. For systems using a common antenna for transmit and receive, each plot is equally valid for jamming and intercept vulnerability evaluations, because plots of power density relative to the power density at boresight, which are used for intercept evaluation, are also a measure of the sensitivity of the radar to jamming noise, relative to the noise power received from a jammer located at boresight intercept.

Normalization of the plots is accomplished by expressing the antenna gain G_t in terms of the peak gain G_m and the relative power gain pattern in azimuth and elevation, $F(A;E)$: $G_t = G_m \times F(A;E)$, and by normalizing the jammer range R_j and the interceptor range R_i to the boresight slant range R_o: $R_{j;i} = (R/R_o)R_o$. The term G_t/R_j^2 in equation (2.1), and G_t/R_i^2 in equation (2.8) can then be expressed by

$$\frac{G_t}{(R_{j;i})^2} = \frac{G_m}{R_o^2} \times \frac{F(A;E)}{(R/R_o)^2} \tag{3.1}$$

The first term is proportional to an absolute value at boresight, and the second term is proportional to values at azimuth angles "A," elevation angles "E," and slant ranges "R," relative to the value at boresight intercept. That term, $F(A;E)/[R/R_o]^2$, is 1, or 0 dB at boresight, and generally lower at off-boresight locations.

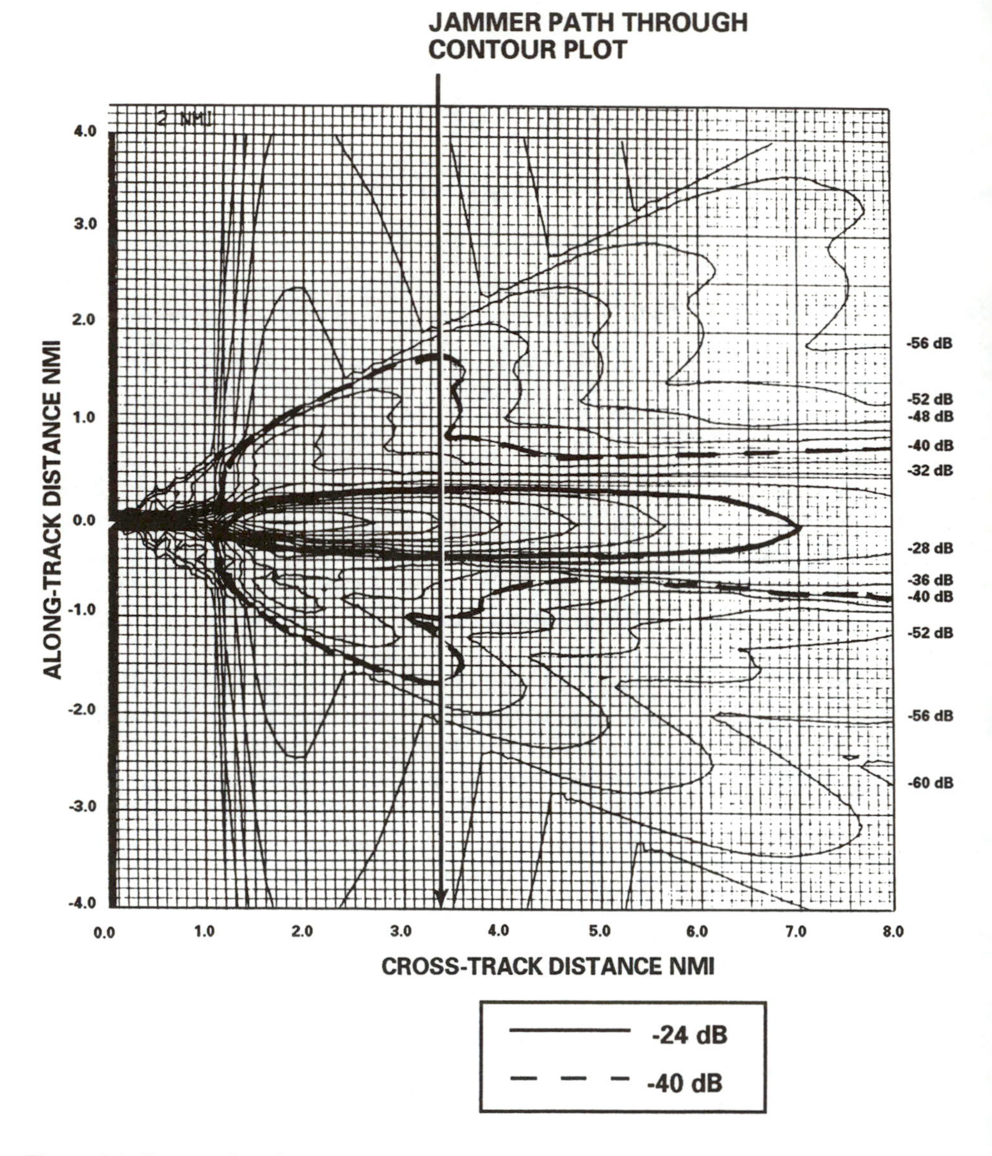

Figure 3.1 Contour plot of nonsymmetrical antenna pattern.

Figures 3.1 and 3.2 are examples of the contour plots. The interval between lines can be selected to be any convenient number, here 4 dB, and the plots are computed for the condition where

$$\frac{F(A;E)}{(R/R_o)^2} = \text{constant } (0 \text{ dB}, -4 \text{ dB, etc.}) \tag{3.2}$$

Figure 3.1 is the contour plot of an antenna with an azimuth beamwidth of approximately 6° and peak azimuth sidelobes of approximately −30 dB. The elevation pattern is shaped over a wider angular range. The boresight axis of the antenna intercepts the ground at a cross range of 2 nmi, and an elevation null at approximately 1 nmi can be observed.

Figure 3.2 is an example of a contour plot of a near-circular antenna pattern, and that pattern is squinted by 45° from the normal to the flight path. The azimuth beamwidth is approximately 18°, and the boresight axis of the antenna intercepts the ground at an along-track and cross-track distance of 1.3 nmi. The peak sidelobes are on the order of −30 dB, and two elevation nulls are displayed.

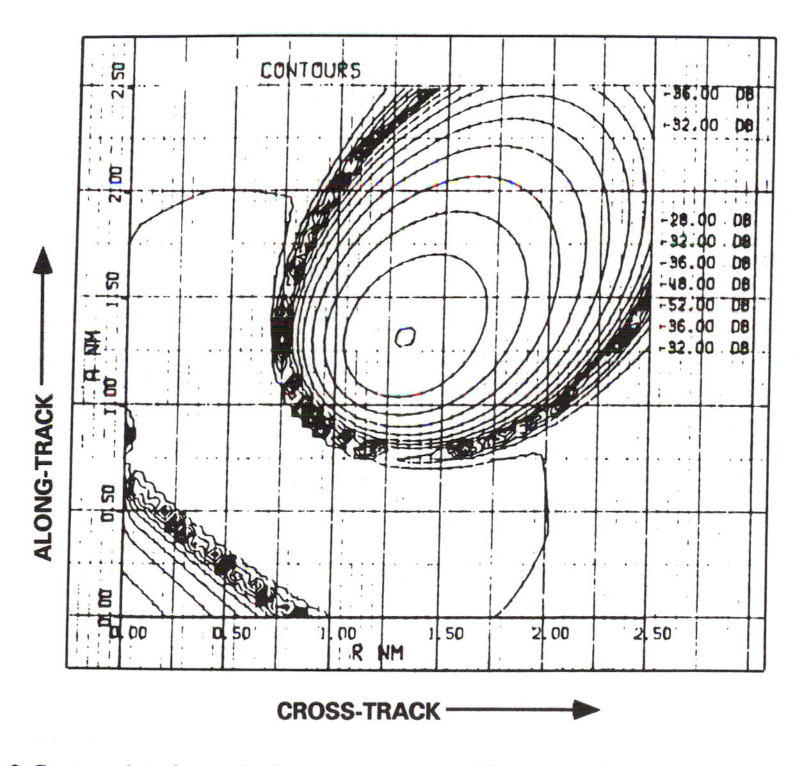

Figure 3.2 Contour plot of near-circular antenna pattern, 45° squint angle.

3.1.2 Intercept Contour Values

The condition for detection of the radar signal was given in equation (2.9). Using that equation and equation (3.1), we can solve for the threshold contour value, here identified as Q_{min}. Detection will occur for all conditions where

$$\frac{F\,(A;E)}{\left(\dfrac{R}{R_o}\right)^2} > Q_{min} = \frac{(S/N)_i}{\dfrac{P_t\,G_m\,A_i}{4\pi R_o^2 k\,T_s\,B_i}\gamma_i} \tag{3.3}$$

The right-hand side of the equation is the ratio of the signal-to-noise ratio $(S/N)_i$, which is required for the detection of the radar signal, to the processed signal-to-noise ratio of the radar signal, if the intercept receiver were located at boresight intercept.

A reduction in the transmitter peak power P_t (such as a quiet mode) results in a new value for the Q_{min}. The reduction in the area enclosed by the contour Q_{min} is a measure of the effectiveness of the quiet-mode operation. Similarly, the use of a low sidelobe antenna can reduce the along-track distance enclosed within lines associated with a particular value. That reduction in along-track distance is again a measure of the effectiveness of the low sidelobe antenna.

3.1.3 Jamming

3.1.3.1 Reconnaissance Objective

In evaluating the effectiveness of jamming, one has to define the reconnaissance objective or, stated differently, where the jammer is effective in denying what kind of information. Two factors have to be considered in stating the reconnaissance objective: the signal-to-noise or clutter-to-noise ratio that the radar will provide in an EW clear environment (that is, the available ratio for the desired targets) and the S/N or C/N ratios needed to extract the desired information.

The available S/N or C/N ratios can be calculated from equations (1.39) or (1.42). But the required S/N ratio or C/N ratio needs to be selected for each reconnaissance task (detect, verify), much like the $(S/N)_i$, had to be selected to meet the probability of intercept and false alarm rate requirement. The required signal-to-noise ratio may be selected to be equal to the design value of the radar for some targets, if the objective is to detect and locate new (unknown) targets. Little degradation from jamming could then be tolerated. On the other hand, if the objective is to verify the continued presence of known targets, then considerable degradation of the image quality from jamming could be tolerated. In the first case the radar is more vulnerable to jamming; in the second case the same radar can be considered tolerant to jamming. We define a jamming tolerance factor K_j as the ratio of the available signal-to-noise ratio for targets with a particular radar cross section to the signal-to-noise ratio that is required to extract the needed information from the radar image:

$$K_j = (S/N)_{\text{available}}/(S/N)_{\text{required}}$$

$$\text{or} \quad K_j = N_{\max}/N_{ra}$$

This is the ratio of the maximum tolerable noise level—which yields the required signal-to-noise ratio—to the processed thermal noise level. Similarly, K_j can be defined for clutter-to-noise ratios, when the reconnaissance objective is related to extended features, and also for signal-to-clutter ratios for those cases in which point target detection in clutter is desired.

The signal-to-noise, clutter-to-noise, and signal-to-clutter ratios required to detect and locate new targets may be on the order of 10 dB and, to verify the continued presence of known targets, perhaps 3 dB.

3.1.3.2 Jamming Contour Value

Jamming will be considered effective when the signal-to-jam ratio drops below the required signal-to-noise ratio, or when the processed jammer noise N_{jra} exceeds the processed thermal noise by the jamming tolerance factor K_j which has to be derived for each target type, mode, and reconnaissance objective.

The incorporation of electronic counter-countermeasures reduces the processed jammer noise. We define an ECCM effectiveness factor K_e, and the condition for successful jamming then becomes

$$N_{jra}/K_e > N_{ra}\,K_j = N_{\max} \tag{3.4}$$

for $N_{jra} \gg N_{ra}$

K_j = the jamming tolerance factor
K_e = the ECCM effectiveness factor

Using equations (2.4) and (3.1), one can solve for the threshold contour value at which jamming becomes effective, which we again identify as $Q_{\min}$. Jamming will be effective for all conditions where

$$\frac{F(A;E)}{(R/R_o)^2} > Q_{\min} = \frac{k\,T_s\,B_r\left(\dfrac{4\pi R_o}{\lambda}\right)^2}{P_j\,G_j\,K_f\,G_m}\,K_j\,K_e \tag{3.5}$$

The right-hand side of the equation states the ratio of the radar's thermal noise to the effective jammer noise, for a jammer located at boresight intercept, in addition to the two factors that have been identified earlier.

A jammer located within the Q_{min} contour lines will be effective against the radar. Incorporation of ECCM devices, such as a notch filter, will reduce the jammer power in the radar system and result in a new contour Q_{min} value. The difference in area enclosed between the contour lines is then the measure of the effectiveness of the ECCM device.

3.2 GENERATING CONTOUR PLOTS

3.2.1 General

Contour plots are unique to mission parameters of a radar (altitude, slant range to boresight, and squint angle), as well as the one-way radar antenna pattern in azimuth and elevation. To generate the plots, a computer program is needed to examine the area of interest surrounding the boresight intercept and calculate the slant range, as well as the azimuth and elevation angles from boresight. The contour term (equation (3.2)) is then formed, and if the value of that term matches a predetermined value (0 dB, −4 dB, and so on), then that point is plotted.

3.2.2 Antenna Patterns

The most accurate values of the antenna patterns of each specific azimuth and elevation angle would be obtained by squaring the Fourier transform of the aperture illumination function, but this process may prove to be too time-consuming. Moreover, the resultant contour plots in the sidelobe region would be too crowded, because this process would result in the definition of the fine structure of the sidelobes.

Adequate results can be obtained by generating look-up tables of the azimuth pattern $F(A)$ and elevation pattern $F(E)$. These patterns describe the mainlobe (and perhaps the first sidelobe of the pattern) in detail and the remaining sidelobes by their envelope, rather than their fine structure.

The value of interest of the antenna pattern for antennas with widely differing azimuth and elevation beamwidths can be determined by forming the product of the pattern values at angles A and E. $F(A;E) = F(A)F(E)$. The contour plot shown in Figure 3.1 is an example for such an antenna. In that example, the mainlobe and the first sidelobe have been defined in detail (evidenced by the existence of the first two sidelobe nulls). Furthermore, the depth of the null was not assigned a very low value, and in the area of interest, the lines can be clearly associated with contour values.

The value of interest for antennas with near-circular patterns can be determined by calculating the off-boresight angle and using the pattern value at that angle. Figure 3.2 is an example of contour plots for antennas with near-circular antenna patterns. In this example, the first sidelobe null was given a value of −60 dB, and the area near the sidelobe null is cluttered. Even though this pattern was defined more accurately, the resultant plot does not offer any more insight into the vulnerability of the radar to intercept and jamming.

3.2.3 Coordinate System

The along-track axis of the plot is the projection of the velocity vector on the ground; the cross-track axis is normal to it. When the radar operates in a side-looking mode, the contour plot is centered on the cross-track axis.

The same plots can be used for a squint mode operation with mechanically squinted antennas. The plot is simply rotated by the squint angle. New patterns have to be generated for each squint angle of electronically squinted antennas, because their pattern changes with the squint angle, and the principal elevation surface becomes a conical half-shell.

3.2.4 Special Case: Jammer Offset Equals Boresight Intercept

Contour plots are universal in that they allow one to evaluate the jamming sensitivity and intercept susceptibility for any offset of the jammer and intercept receiver from the flight path, within the size of the plot. Generation of such a plot requires, at least initially, preparation of a computer program.

To obtain a quick assessment of the EW aspects of a radar for the special case that the jammer (or intercept receiver) offset from the flightpath coincides with the boresight intercept of the radar antenna on the ground ($R_{i;j} = R_o$), one can generate a plot of contour values for that specific case only. This has been done for an antenna with a 1.43° azimuth beamwidth and a uniform aperture illumination function, which results in a $(\sin x/x)^2$ shaped pattern. The azimuth pattern through the boresight axis (the principal azimuth plane) is then $F(A;0) = (\sin x/x)^2$, with $x = \pi\,(L/\lambda)\sin A$, where A is the azimuth angle off boresight. The contribution of the change in range to the contour plot term becomes $(R_o/R)^2 = (\cos A)^2$.

In the example of Figure 3.3, the mainlobe and the fine structure of the sidelobes are omitted and the sidelobes are defined by their envelope, at 3 dB below their peak values. The sidelobes decrease with $1/x^2$, and the range dependency of the contour values becomes noticeable for azimuth angles greater than 15°. This plot can now be used for any radar altitude and beam elevation angle, but the jammer offsets have to be equal to the boresight intercept. Figure 3.3 shows the contour

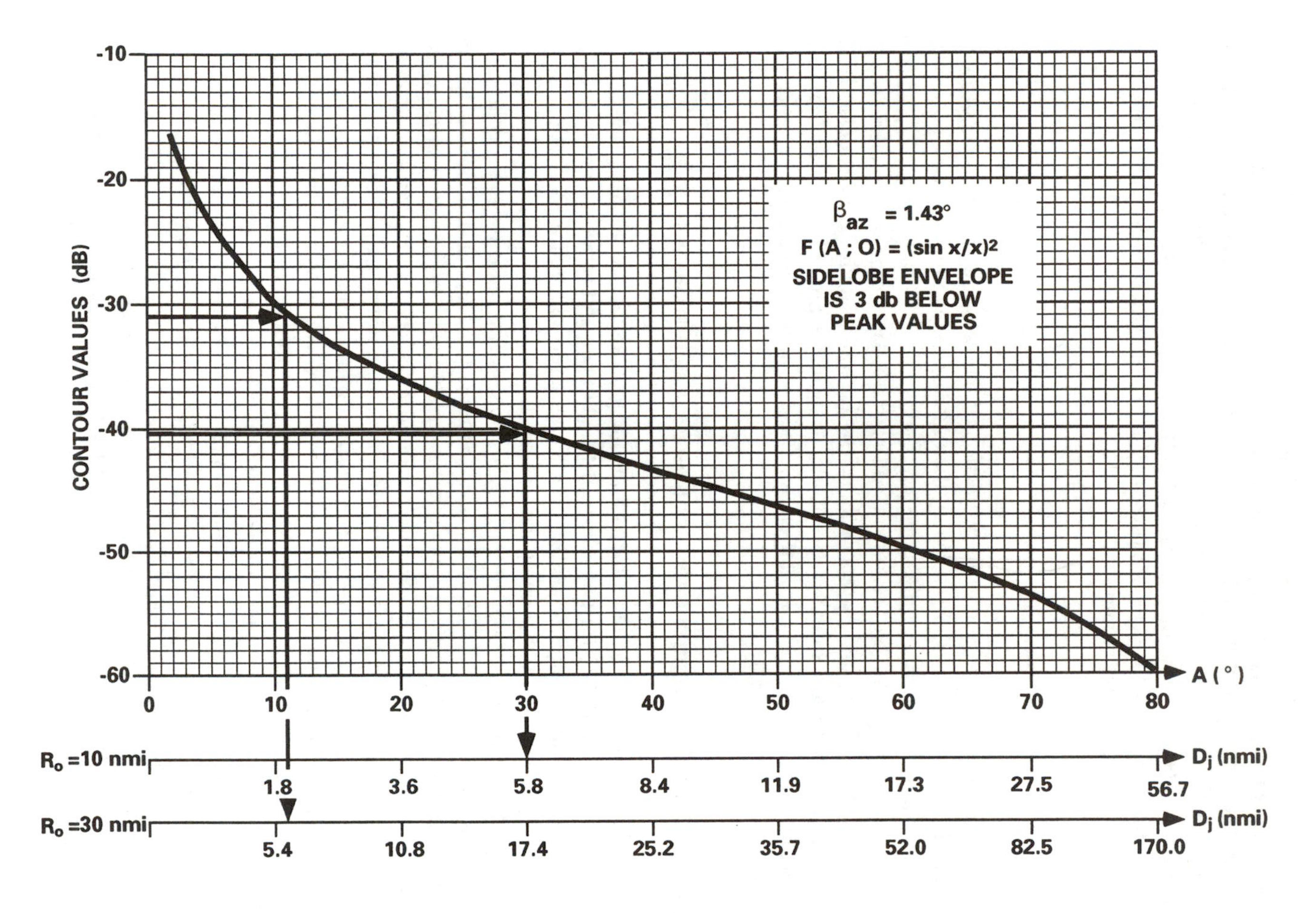

Figure 3.3 Contour values versus azimuth angle (A), and along-track distance D_{ji}, R_{ji} = R_o.

values as a function of the azimuth angle A from 0° to 80°. The corresponding along-track distance for any contour value is shown for boresight distances of 10 and 30 nmi. The total distance over which a jammer or interceptor is effective is twice the value so obtained.

3.3 EVALUATION PROCEDURE

The procedure for evaluating the susceptibility of a sensor to ECM includes the following:

To assess the vulnerability of the radar to jamming,

- Define the mission objective in terms of the available and required signal-to-noise ratio (that is, define K_j).
- Using equation (3.5), calculate Q_{min}, the contour value at which the jammer is effective.
- Generate the contour plot for the given mode, altitude, and pattern, and annotate the line associated with the calculated Q_{min} value. The area enclosed within this contour defines the location of the jammer relative to the location and flight path of the sensor carrier, where the jammer is effective against the sensor.
- Iterate this process for different ECCM fixes and evaluate the effectiveness of the ECCM devices by noting the decrease in the area enclosed.

The procedure for assessing the vulnerability to intercept is similar, except that equation (3.3) is used to calculate Q_{min} and the area enclosed within the contour lines defines the location of the intercept receiver at which the radar is detected. Furthermore, the process has to be iterated for different peak power levels (quiet mode) and different LPI waveforms.

In the example in Figure 3.1, the area enclosed within the contour line associated with -24 dB has been enhanced by a solid line. If equation (3.3) had resulted in $Q_{min} = -24$ dB, then an intercept receiver located anywhere within this area would detect the presence of the radar, while the radar was imaging an object at zero along-track distance. One can visualize the process of detection by sliding the contour lines along a flightpath. When the leading edge of the contour Q_{min} crosses the location of an intercept receiver, then the presence of the radar is detected and will remain detected until the trailing edge of that line crosses the location of the interceptor. If equation (3.5) had resulted in $Q_{min} = -24$ dB, then a jammer located anywhere within this area would be effective in denying the mission objective of imaging targets at that along track. The jammer would continue to be effective against such targets for a flightpath equal to the along-track distance within the contour line. The example that was chosen shows a jammer or interceptor to be effective only in the mainlobe of the radar antenna. Had the equations yielded values for Q_{min} that were less than -40 dB (dotted line in Figure 3.1), detection and effective jamming would have occurred in the sidelobe region, thereby greatly enhancing the effectiveness of jamming.

Imaging radars capable of operating in a number of modes, which differ primarily in their maximum imaging range but can also differ in altitude, squint angle, and resolution, are designed to provide "acceptable" or "specified" S/N and C/N ratios in their longest-range modes and result in significantly higher values in their shorter-range modes. It will generally be found that such a radar, when imaging in its long-range mode, would be vulnerable to jamming before it is actually detected by the intercept receiver (that is, the radar could be jammed if its presence were known). Conversely, when the radar operates in its short-range mode, the radar will be visible to an intercept receiver before the jammer noise is strong enough to obscure an image. Without power management, such radars can be said to operate in a quiet mode in their long-range modes and provide burnthrough in their short-range modes. Because both intercept and jamming must be successful to deny radar reconnaissance, the region of vulnerability is the smaller of the two areas defined by (3.3) and (3.5).

3.4 SIMULATED JAMMER NOISE AND EVALUATION

Figure 3.4 compares the radar image in an EW clear environment with one corrupted by jammer noise. The radar image, courtesy of Loral Defense Systems, Arizona, shows the area near Sun City, Arizona, and was selected because it shows strong point targets, such as electric power line towers, as well as low-reflectivity agricultural patterns and desert land. The jammer noise has been simulated, and the intensity of the noise has been selected to demonstrate the effectiveness of jamming against different targets.

3.4.1 Geometric Evaluation

The jammer was in the principal elevation plane (the elevation plane through the boresight axis) of the radar where the jammer noise is a maximum. It may be difficult to identify that maximum, thus a more accurate method of determining the along-track location of the jammer is to select the line halfway between clearly identifiable sidelobe nulls. The cross-track location of the jammer can now be determined from the along-track distance between those nulls and the knowledge of the angle between the sidelobe nulls of the azimuth pattern of the radar antenna.

In the example, the first sidelobe null was not shown to be clear of jammer noise because the jammer noise intensity at any along-track location is not only governed by the gain of the antenna in that direction, but also by the sum of the jammer noise from an angular region associated with one synthetic-aperture length. That angle, which can be several tenths of a degree, will include data from the mainlobe

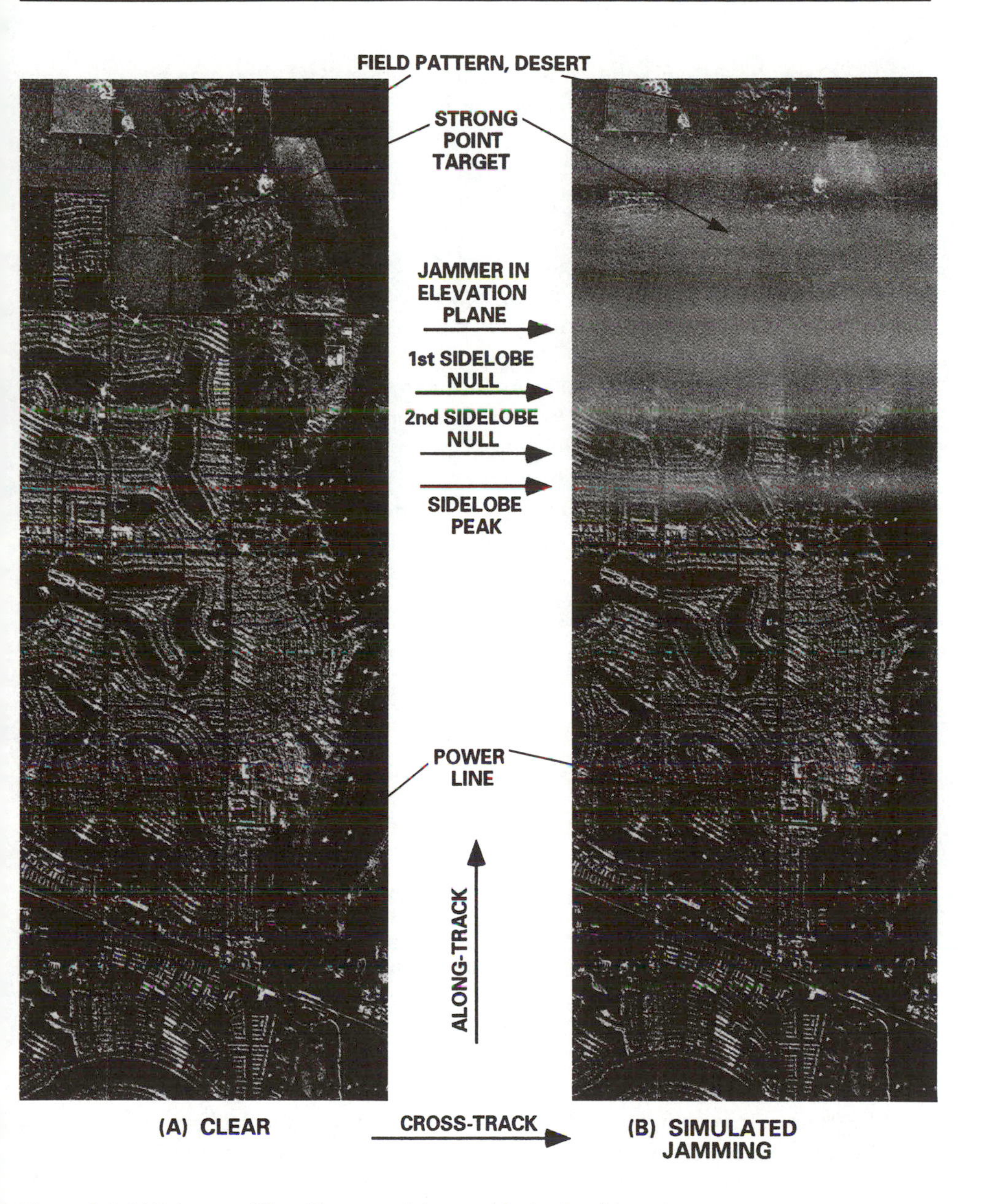

Figure 3.4 SAR image of Sun City area, Arizona with simulated jamming.

and the first sidelobe, in addition to data from the null. (This is another reason the sidelobe nulls of the antenna patterns need not be defined in great detail.)

3.4.2 Scatterometric Evaluation

The jammer noise has been simulated, and therefore that radar image cannot be accepted as proof of statements made earlier about reconnaissance objectives and jammer effectiveness. Rather, the intent is to convey a general impression of jammed imagery and to show trends. Figure 3.4 is annotated to identify field patterns and desert areas, an isolated strong point target, and an electrical power line extending through the entire along-track dimension of the radar image. Jammer noise is shown to block all radar data in the mainlobe of the radar antenna, partially obscuring radar data in the first and second sidelobes and having no effect in the second sidelobe null. This effect was created by superimposing a jammer noise pattern on the radar image. The jammed image shows that the electrical power line can be detected (or verified) everywhere except in the mainlobe. The strong isolated point target is clearly visible through the jammer noise in the first sidelobe peak; the details of the desert and field patterns tend to be degraded.

3.5 NUMERICAL EXAMPLE

To illustrate the susceptibility of the radar to intercept, and the vulnerability to jamming, we use the example of the AN/APQ-102A radar of Section 1.4 and postulate intercept and jammer parameters.

3.5.1 Susceptibility to Intercept

From equation (3.3) we find that the value Q_{min} associated with the 10-nmi mode is -40 dB and -31 dB for the 30-nmi mode (approximate values). The radar would be detected through the sidelobes of the radar antenna, because the radar antenna peak sidelobes are only 13.4 dB below the value at boresight.

To reduce the along-track distance enclosed within the intercept contour value, one can apply RF power management and use a low sidelobe antenna. It is observed that the radar yields a higher S/N ratio and C/N ratio in the 10-nmi mode than in the 30-nmi mode. If the performance at 30 nmi is acceptable, then the peak power in the 10-nmi mode could be reduced by approximately 14 dB, with no loss in acceptable radar performance. The Q_{min} value associated with this new power level is now -26 dB. In a quiet mode, such as one transmitting only enough RF power to image 100 ft^2 targets with a S/N ratio of 10 dB, the radar peak power could be reduced further and the resultant Q_{min} values would be well into the region of the main lobe, namely -5.6 dB at 10 nmi and -10.5 dB at 30 nmi (see Table 3.2).

A low sidelobe antenna with peak sidelobes of −30 dB would confine the detection of the radar to the mainlobe region in the 30-nmi mode. It would require a low sidelobe antenna and some power management to achieve the same goal in the 10-nmi mode. The radar would then be observed by the intercept station while the radar antenna mainbeam sweeps over the intercept station site; that distance is 1520 ft at 10 nmi and 4560 ft at 30 nmi. At 900 ft/sec the radar would remain detected for 1.7 and 5 seconds in the two modes, respectively.

3.5.2 Vulnerability to Jamming

The vulnerability evaluation will be made for three different target types and two different objectives. The following K_j values result from using the values derived in Section 1.4 and listed in Table 3.1 for a 100 ft^2 point target, area targets with a radar backscatter coefficient of −20 dB, and also the signal-to-clutter ratio for the above target types, at maximum ranges of 10 and 30 nmi, as well as a required ratio

Table 3.1
Radar and EW Parameters

Parameter	*Mode*		*Remark*
	10 nmi	*30 nmi*	
RADAR			
RF peak power (kW)	50		
Antenna gain (dB)	30		
Azimuth beamwidth (deg)	1.43		
Azimuth peak sidelobes (dB)	−13.4		Uniform illumination
Signal to noise σ = 100 ft^2 (dB)	44.7	30.4	From Table 1.6
Clutter to noise σ_0 = −20 dB (dB)	34.7	20.4	From Table 1.6
Signal to clutter (dB)	10.0	10.0	From Table 1.6
INTERCEPT STATION			
Thermal noise (dBm)	−75		Postulated values
Required S/N_i (dB)	14		Postulated values
Antenna gain G_i (dB)	10		Postulated values
γ_i (dB)	0		Postulated values
Range to intercept Station R_i (nmi)	10	30	
JAMMER			
Peak ERP (kW)	100		Postulated values
Bandwidth (MHz)			
Barrage	1500		Postulated values
Spot, pulse	5		Postulated values
Duty cycle	0.1		Postulated values
Range to R_j jammer (nmi)	10	30	

of 10 dB for the detection of unknown targets, and 3 dB for the verification of known targets.

We find that the available S/N and C/N ratios exceed the required ones by a significant margin and conclude that the radar is tolerant to jamming for those objectives (large K_j values).

Whereas the noise floor (see Section 2.2.2.3) for the above cases was thermal noise, we find that when the objective is to detect point targets in clutter, the noise floor is clutter. Jammer noise adds to that noise floor, and to determine the tolerance to jamming one has to form the signal-to-(clutter plus processed jammer noise) ratio. Table 3.3 shows that the available S/C ratio marginally meets the required ratio for detection, and it would appear that no additional jammer noise can be tolerated. However, if we allow the jammer noise to degrade the S/C ratio by a modest 1 dB, as shown in Table 3.3, then the tolerance to jamming for this case ranks between that of detecting point targets and that of detecting extended features.

Table 3.2

Intercept Contour Value Q_{min} Summary

Condition	Mode	
	10 nmi	*30 nmi*
Without ECCM (dB)	−40	−31
14-dB power reduction in 10-nmi mode (dB)	−26	
Quiet mode: S/N = 10 dB for σ = 100 ft^2 (dB)	−5.6	−10.5

Table 3.3

Available S/N, C/N, and S/C Ratios, Required Ratios, and K_j Values (in dB)

Reconnaissance Objective	R_o = *10 nmi*			R_o = *30 nmi*		
	Available	*Required*	K_j	*Available*	*Required*	K_j
S/N for σ = 100 ft^2						
Detect	44.7	10	34.7	30.4	10	20.4
Verify	44.7	3	41.7	30.4	3	27.4
C/N for σ_o = −20 dB						
Detect	34.7	10	24.7	20.4	10	10.4
Verify	34.7	3	31.7	20.4	3	17.4
S/C for σ = 100 ft^2, σ_o = −20 dB						
Detect	10.0	10	29.0*	9.9	10	14.5*
Verify	10.0	3	40.7	9.9	3	26.4

*Lowers the S/C ratio by 1 dB

To derive the contour values at which jamming is effective in denying the stated objectives, we calculate the contour term from equation (3.5); the spectral mismatch factor K_j is 0 dB for the spot jammer and the random pulse jammer case and −20 dB for the barrage jammer mode. The effective ERP for the spot and barrage jammer is the peak ERP, and in the random pulse mode is the average transmitted ERP. With a duty cycle of 10%, that value is 40 dBW. K_e is 0 dB for all cases, because ECCM is not being evaluated at this time.

The contour value Q_{min}, exclusive of the jamming tolerance factor K_j (equation (3.5)), is then −69.5 dB for the spot jammer, −49.5 dB for the barrage jammer, and −59.5 dB for the random pulse jammer, in the 10-nmi range mode. The corresponding values in the 30-nmi mode are −60, −40, and −50 dB.

It is observed that in the shorter-range mode (10 nmi), the presence of the radar is detected before the jammer can be effective—a form of burnthrough. In the longer-range mode the jammer could be effective over a longer along-track distance if the presence of the radar were known—a form of quiet mode operation.

3.5.3 Along-Track Distance of Effective Jamming

Figure 3.3 has been annotated to show the along-track distances associated with the intercept Q_{min} values. The −40 dB intercept contour threshold value corresponds to an azimuth angle of ±30°, which results in an along-track distance at R_o = 10 nmi of ±5.8 nmi. The −31 dB intercept contour threshold value is associated with an azimuth angle of ±11°, which also results in an along-track distance of ±5.8 nmi at R_o = 30 nmi. Other Q_{min} values for specific reconnaissance objectives can be translated to along-track distances in the same manner.

3.5.4 ECCM

To defeat the jammer, the radar would need to have built-in ECCM features that have an effectiveness factor equal to the reciprocal (positive dB value) of the Q_{min} value that is associated with each particular reconnaissance objective. In reviewing the values of Tables 3.4 and 3.5, one concludes that ECCM features that reduce the jammer noise by 20 to 30 dB would defeat the barrage jammer. The same ECCM features would confine the effectiveness of the spot and randomly pulsed jammer to the mainlobe of the radar's azimuth antenna pattern.

3.5.5 Burst Mode (see also Section 2.4.2.3)

In the burst mode, a radar is operated only long enough to generate the synthetic-aperture required to process the radar data at the maximum range to the desired azimuth resolution. In addition to the targets centered on 0-doppler, all targets within

Table 3.4
Contour Values $Q_{\min}$ in dB, R_o = 10 nmi

Reconnaissance Objective	*Jammer Type/ERP* Spot 50 dBW	Barrage 30 dBW	Pulsed 40 dBW
S/N, Detect	−34.7	−14.7	−24.7
Verify	−27.8	−7.8	−17.8
C/N, Detect	−44.5	−24.5	−34.5
Verify	−37.5	−17.5	−27.5
S/C, Detect	−40.5	−20.5	−30.5
Verify	−28.8	−8.8	−18.8
Intercept	−40	−40	−40

Table 3.5
Contour Values $Q_{\min}$ in dB, R_o = 30 nmi

Reconnaissance Objective	*Jammer Type/ERP* Spot 50 dBW	Barrage 30 dBW	Pulsed 40 dBW
S/N, Detect	−39.5	−19.5	−29.5
Verify	−32.5	−12.5	−22.5
C/N, Detect	−49.6	−29.6	−39.6
Verify	−42.6	−22.6	−32.6
S/C, Detect	−45.5	−25.5	−35.5
Verify	−33.6	−13.6	−23.6
Intercept	−31	−31	−31

the mainlobe of the azimuth antenna pattern (except those at the trailing edge of the antenna pattern) remain illuminated for one synthetic-aperture length, and their doppler history is centered on doppler frequencies other than 0. The radar returns from those targets are then processed (in the batch mode rather than a line-by-line mode) to generate additional range lines, and the processor generates an entire scene from the radar data collected during one short burst of radar signals. The radar is off during the time required to fly the length of that scene, after which another burst is transmitted to generate the adjacent scene.

Assume that the AN/APQ-102A radar were operated in a burst mode (which it was not designed for) and further assume the following numerical values:

Maximum range R_{max} = 30 nmi
Minimum range R_{min} = 20 nmi
Processed azimuth resolution w_a = 30 ft
Aircraft velocity v_a = 900 ft/sec

The synthetic-aperture length for a 30-ft resolution at a slant range of 30 nmi is 304 ft, and the radar on-time is then $t_s = 0.34$ sec.

The radar can then be turned off for the time required to fly the maximum scene size, which is the maximum synthetic-aperture length at the minimum range (equation 1.19) minus the synthetic-aperture length at the maximum range (equation 1.21). In this example the scene size is 3040 ft −304 ft = 2736 ft, and the off-time is then 2.7 seconds.

If the intercept station were located 30 nmi from the radar's flight path, then the intercept station could detect the presence of the radar over a flight path of 4500 ft (mainlobe intercept) and maintain detection for 5 seconds if operated in the continuous mode of operation. In the burst mode, however, a maximum of 2 bursts of 0.34 seconds each would be observed, and the radar would be off for the remainder of that flight path. When combined with frequency diversity and low sidelobe antennas, the jammer would have less opportunity to initiate effective jamming. Additionally, the radar could transmit deception signals between bursts, when the radar is not needed to collect radar data.

List of Symbols

A_r	=	Effective radar receiving antenna area
A^2	=	Processing gain for uniformly weighted spectrum
A_w^2	=	Processing gain for weighted spectrum
B_d	=	Processed doppler bandwidth
B_i	=	Instantaneous intercept receiver bandwidth
B_j	=	Jammer noise bandwidth
B_n	=	Effective reject bandwidth of notch filter
B_r	=	Radar signal bandwidth
c	=	Speed of light
C	=	Total power of aperture
C_1	=	Clutter return from one pulse
C_{ra}	=	Clutter level after range and azimuth processing
$(C/N)_1$	=	Clutter to thermal noise from one pulse
$(C/N)_{ra}$	=	Clutter to thermal noise after range and azimuth processing
D_j	=	Jammed along-track distance
$F(A;E)$	=	Radar antenna pattern in azimuth and elevation
f_j	=	Jammer RF center frequency
f_r	=	Radar RF center frequency
G_i	=	Intercept antenna gain in the direction of the radar
G_j	=	Jammer antenna gain in the direction of the radar
G_t	=	Radar transmit antenna gain
G_m	=	Radar antenna peak gain
k	=	Boltzmann's constant, 1.38×10^{-23} watts/(Hertz $\times$ kelvins)
K_a	=	Azimuth oversampling factor
K_e	=	ECCM improvement factor
K_f	=	Spectral mismatch factor, $0 \leq K_f \leq 1$
K_j	=	Jamming tolerance factor
L	=	Length of physical antenna

L_m = Reduction in signal-to-noise due to weighting
L_s = Length of synthetic antenna
n = Number of azimuth samples coherently processed
n_c = Number of azimuth samples collected (in one synthetic-aperture length)
n_r = Number of phase code segments in transmit pulse
N = Thermal noise
NF = Noise figure
N_1 = Thermal noise from one pulse
N_{ra} = Thermal noise after range and azimuth processing
N_i = Thermal noise of intercept receiver
N_j = Effective jammer noise
N_{ji} = Jammer noise intercepted by radar
N_{jra} = Jammer noise after range and azimuth processing
PRF = Pulse repetition frequency
$\mathrm{PRF}_{\min}$ = Minimum transmitted PRF
P_j = Peak jammer transmitted power
P_t = Peak radar transmitted power
P_{jr} = Jammer power intercepted by radar
P_{ri} = Radar power intercepted by intercept receiver
$Q_{\min}$ = Intercept or jamming threshold value
R = Slant range to target
R_i = Distance between radar and intercept antenna
R_j = Distance between radar and jammer
R_0 = Slant range from radar to boresight intercept on the ground
S_1 = Radar signal return from one pulse
S_r = Radar signal power after range processing
S_{ra} = Radar signal strength after range and azimuth processing
$(S/N)_1$ = Signal-to-noise ratio from one pulse
$(S/N)_i$ = Signal-to-noise ratio required by intercept receiver
$(S/N)_{ra}$ = Signal-to-noise ratio after range and azimuth processing
$(S/N_j)_{ra}$ = Signal-to-jammer noise ratio after range and azimuth processing
T_s = System noise temperature in kelvins
t_s = Synthetic-aperture time
v_a = Velocity of radar carrier
w_a = Processed cross-range resolution
$(w_a)_b$ = Best azimuth resolution in a strip mapping mode
w_r = Processed range resolution
W_a = Real aperture cross-range resolution
W_r = Unprocessed range resolution
β_{null} = Null angle of physical antenna

β_{az}	=	3-dB azimuth beamwidth of physical antenna
$(\beta_{null})_s$	=	Null angle of synthetic antenna
$(\beta_{az})_s$	=	3-dB beamwidth of synthetic antenna
η	=	Aperture efficiency factor
η_a	=	Azimuth aperture efficiency factor
η_r	=	Range aperture efficiency factor
γ_a	=	Azimuth compression ratio
γ_r	=	Range compression ratio
γ_i	=	Intercept receiver processing gain
θ_g	=	Grazing angle
λ	=	Wavelength
σ	=	Radar cross section of point target
σ_0	=	Clutter backscatter coefficient
τ_c	=	Processed pulsewidth
τ_t	=	Transmitted pulsewidth

Index

The Artech House Radar Library

David K. Barton, Series Editor

Active Radar Electronic Countermeasures, Edward J. Chrzanowski

Adaptive Signal Processing for Radar, Ramon Nitzberg

Airborne Early Warning Radar, William Morchin

Airborne Early Warning System Concepts, Maurice Long, et al.

Airborne Pulsed Doppler Radar, Guy V. Morris

AIRCOVER: Airborne Radar Vertical Coverage Calculation Software and User's Manual, William A. Skillman

Analog Automatic Control Loops in Radar and EW, Richard S. Hughes

Aspects of Modern Radar, Eli Brookner, et al.

Aspects of Radar Signal Processing, Bernard Lewis, Frank Kretschmer, and Wesley Shelton

Bistatic Radar, Nicholas J. Willis

CARPET (Computer-Aided Radar Performance Evaluation Tool): Radar Performance Analysis Software and User's Manual, developed at the TNO Physics and Electronic Laboratory by Albert G. Huizing and Arne Theil

Cryptology Yesterday, Today, and Tomorrow, Cipher Deavours et al.

Cryptology: Machines, History, and Methods, Cipher Deavours et al.

DETPROB: Probability of Detection Calculation Software and User's Manual, William A. Skillman

Designing Automatic Control Systems, W.L. Rubin

Detectability of Spread-Spectrum Signals, Robin A. Dillard and George M. Dillard

Electronic Countermeasure Systems Design, Richard Wiegand

Electronic Homing Systems, M.V. Maksimov and G.I. Gorgonov

Electronic Intelligence: The Analysis of Radar Signals, Richard G. Wiley

Electronic Intelligence: The Interception of Radar Signals, Richard Wiley

Engineer's Refractive Effects Prediction Systems–EREPS, Herbert Hitney

EREPS: Engineer's Refractive Effects Prediction System Software and User's Manual, developed by NOSC

High Resolution Radar, Donald R. Wehner

High Resolution Radar Cross-Section Imaging, Dean Mensa

Interference Suppression Techniques for Microwave Antennas and Transmitters, Ernest R. Freeman

Introduction to Electronic Defense Systems, Fillippo Neri

Introduction to Electronic Warfare, D. Curtis Schleher

Introduction to Sensor Systems, S.A. Hovanessian

IONOPROP: Ionospheric Propagation Assessment Software and Documentation, Herbert Hitney

Kalman-Bucy Filters, Karl Brammer

Laser Radar Systems, Albert V. Jelalian

Lidar in Turbulent Atmosphere, V.A. Banakh and V.L. Mironov

Logarithmic Amplification, Richard Smith Hughes

Machine Cryptography and Modern Cryptanalysis, Cipher A. Deavours and Louis Kruh

Millimeter-Wave Radar Clutter, Nicholas Currie, Robert Hayes, and Robert Trebits

Modern Radar System Analysis, David K. Barton

Modern Radar System Analysis Software and User's Manual, David K. Barton and William F. Barton

Monopulse Radar, A.I. Leonov and K.I. Fomichev

MTI and Pulsed Doppler Radar, D. Curtis Schleher

Multifunction Array Radar Design, Dale R. Billetter

Multisensor Data Fusion, Edward L. Waltz and James Llinas

Multiple-Target Tracking with Radar Applications, Samuel S. Blackman

Multitarget-Multisensor Tracking: Advanced Applications, Volume I, Yaakov Bar-Shalom, ed.

Multitarget-Multisensor Tracking: Advanced Applications, Volume II, Yaakov Bar-Shalom, ed.

Over-The-Horizon Radar, A.A. Kolosov, et al.

Principles and Applications of Millimeter-Wave Radar, Charles E. Brown and Nicholas C. Currie, eds.

Pulse Train Analysis Using Personal Computers, Richard G. Wiley and Michael B. Szymanski

Radar and the Atmopshere, Alfred J. Bogush, Jr.

Radar Anti-Jamming Techniques, M.V. Maksimov, et al.

Radar Cross Section Analysis and Control, A.K. Bhattacharyya and D.L. Sengupta

Radar Cross Section, Eugene F. Knott, et al.

Radar Electronic Countermeasures System Design, Richard J. Wiegand

Radar Engineer's Sourcebook, William Morchin

Radar Evaluation Handbook, David K. Barton, et al.

Radar Evaluation Software, David K. Barton and William F. Barton

Radar Meteorology, Henri Sauvageot

Radar Polarimetry for Geoscience Applicatons, Fawwaz Ulaby

Radar Propagation at Low Altitudes, M.L. Meeks

Radar Reflectivity Measurement: Techniques and Applications, Nicholas C. Currie, ed.

Radar System Design and Analysis, S.A. Hovanessian

Radar Technology, Eli Brookner, ed.

Radar Vulnerability to Jamming, Robert N. Lothes, Michael B. Szymanski, and Richard G. Wiley

RGCALC: Radar Range Detection Software and User's Manual, John E. Fieldng and Gary D. Reynolds

SACALC: Signal Analysis Software and User's Guide, William T. Hardy

Secondary Surveillance Radar, Michael C. Stevens

SIGCLUT: Surface andVolumetric Clutter-to-Noise, Jammer and Target Signal-to-Noise RadarCalculation Software and User's Manual, William Skillman

Signal Detection and Estimation, Mourad Barkat

Small-Aperture Radio Direction Finding, Herndon Jenkins

Solid-State Radar Transmitters, Edward D. Ostroff, et al.

Space-Based Radar Handbook, Leopold J. Cantafio, ed.

Spaceborne Weather Radar, Robert M. Meneghini and Toshiaki Kozu

Statistical Signal Characterization, Herbert Hirsch

Statistical Signal Characterization: Algorithms and Analysis Programs Software and User's Manual, Herbert Hirsch

Statistical Theory of Extended Radar Targets, R.V. Ostrovityanov and F.A. Basalov

Surface-Based Air Defense System Analysis, Robert Macfadzean

Surface-Based Air Defense System Analysis Software and User's Manual, Robert Macfadzean and James M. Johnson

Synthetic-Aperture Radar and Electronic Warfare, Walter W. Goj

VCCALC: Vertical Coverage Plotting Software and User's Manual, John E. Fielding and Gary D. Reynolds